African Meteorites

Edited by

Abderrahmane Ibhi

Geoheritage and Geomaterials Laboratory
University Museum of Meteorites
Ibn Zohr University, Agadir, Morocco

Giorgio S. Senesi

CNR - Institute for Plasma Science and
Technology (ISTP) - Bari Seat
Bari, Italy

Lahcen Ouknine

Geoheritage and Geomaterials Laboratory
University Museum of Meteorites
Ibn Zohr University, Agadir, Morocco

&

Fouad Khiri

Geoheritage and Geomaterials Laboratory
University Museum of Meteorites
Ibn Zohr University, Agadir, Morocco

Regional Center of Trades of Education Training Inzegane
Agadir, Morocco

African Meteorites

Editors: Abderrahmane Ibhi, Giorgio S. Senesi, Lahcen Ouknine and Fouad Khiri

ISBN (Online): 978-981-5136-29-6

ISBN (Print): 978-981-5136-30-2

ISBN (Paperback): 978-981-5136-31-9

First published in 2023.

need for a court order if at any point you breach any terms of this License Agreement. In no event will any delay or failure by Bentham Science Publishers in enforcing your compliance with this License Agreement constitute a waiver of any of its rights.

3. You acknowledge that you have read this License Agreement, and agree to be bound by its terms and conditions. To the extent that any other terms and conditions presented on any website of Bentham Science Publishers conflict with, or are inconsistent with, the terms and conditions set out in this License Agreement, you acknowledge that the terms and conditions set out in this License Agreement shall prevail.

Bentham Science Publishers Pte. Ltd.
80 Robinson Road #02-00
Singapore 068898
Singapore
Email: subscriptions@benthamscience.net

BENTHAM SCIENCE

CONTENTS

FOREWORD

Meteorites recovered in Africa represent more than 1/6 of all extraterrestrial rocks collected in the entire world. This means that Africa hosts the second-largest population of meteorites after Antarctica. The present book offers complete information on the origin and characteristics of meteorite falls and finds in Africa.

The book opens with an exhaustive synthesis of the formation and origin of meteorites, their categorization into "falls" and "finds" according to the circumstances of their discovery, the criteria for their identification, and the guidelines adopted for the nomenclature of meteorites and their classification. The second chapter deals with a statistical study of falls recorded in different African countries. It analyzes their evolution in time and space, provides information on the distribution of their masses, and characterizes their typological classification, focusing on the different factors likely to contribute to their detection on the ground and recovery. The following chapter is devoted to meteorite finds in Africa, pointing to the advantageous natural and human factors that enhance the discovery of new meteorites on this continent. Again, the distribution of meteorite masses collected and their typological classification are presented, drawing the reader's attention to the influence of climate, terrestrial age, as well as sample porosity and mass, and the variation of the degree of weathering.

The subsequent chapters are devoted to other important aspects involving the study of African meteorites, such as the complex problematics connected to the nomenclature of North West African meteorites (NWA), therefore enhancing the scientific importance of the meteoritic heritage of northwest African countries and, consequently by providing detailed documentation on the circumstances of their recovery. A specific chapter is dedicated to Moroccan meteorites, as this region has proved to be one of the most prolific areas in the world for meteorite recovery. The following chapter deals with the Saharan meteorites, stressing the hypothesis that many new meteorites might be extracted from the Sahara Desert in the coming decades, as supported by the estimate that more than 90% of the desert surface has not yet been explored. The last chapter reports on the North East African meteorites and contains an up-to-date review of the confirmed and proposed meteorite falls and finds and of the possible related impact structures in Egypt, Sudan and Libya. The book ends with a summary of the most significant scientific results obtained from the various studies carried out on African meteorites.

I am honored to present this book that aims to become a reference text for university students, researchers, collectors, meteorite enthusiasts, museum curators, astronomers, and for all those readers interested in African meteorites and, more generally, in celestial rocks.

Finally, I would like to congratulate the editors, the authors and their collaborators, all with long-standing professional activity, for their contribution to the publishing of this comprehensive and well-balanced book on the extraordinary scientific heritage represented by African meteorites.

Vanni Moggi Cecchi
Università Degli Studi Di Firenze
Museo di Storia Naturale – Sistema Museale di Ateneo
Via La Pira, 4 50132 Firenze
Italy

PREFACE

According to Alain Carion, "At the beginning of the 20th century, in a book by the Marquis de Mauroy, he wrote that to discover meteorites, one had to read books, see, and possibly touch real meteorites. The authors, with their books on meteorites and the University Museum of Meteorites of Agadir created by some of them, are in line with the discoverers of knowledge". This is the sense that this book aims to deal with meteorites in Africa. The book that the reader has in his hands is a collection of works performed on meteorites in Africa, where arid and hot zones represent 60% of the continent's surface.

This book is a single item of reference for all researchers who require complete information on the origin and characteristics of meteorite falls and finds in Africa, which includes the second-largest population of meteorites on the Earth. This book also provides unique information related to meteorite classification, and can be an important reference on the favorable factors to collect meteorites in Africa, which still hides a huge number of undiscovered meteorites.

The book will be useful for researchers, meteorite hunters, meteorite enthusiasts, museums, astronomers, students, and any person interested in geology and astronomy.

As it is a voluminous subject, we might have missed some references. Readers are requested to bring such omissions to our notice to be included in future editions.

We thank Dr. Vanni Moggi Cecchi for writing the foreword of this book.

Abderrahmane Ibhi
Geoheritage and Geomaterials Laboratory
University Museum of Meteorites
Ibn Zohr University, Agadir
Morocco

Giorgio S. Senesi
CNR - Institute for Plasma Science
and Technology (ISTP) - Bari Seat
Bari, Italy

Lahcen Ouknine
Geoheritage and Geomaterials Laboratory
University Museum of Meteorites
Ibn Zohr University, Agadir, Morocco

&
Fouad Khiri
Geoheritage and Geomaterials Laboratory
University Museum of Meteorites
Ibn Zohr University, Agadir, Morocco

Regional Center of Trades of Education Training Inzegane
Agadir, Morocco

List of Contributors

Abderrahmane Ibhi — Geoheritage and Geomaterials Laboratory, University Museum of Meteorites, Ibn Zohr University, Agadir, Morocco

Fouad Khiri — Geoheritage and Geomaterials Laboratory, University Museum of Meteorites, Ibn Zohr University, Agadir, Morocco
Regional Center of Trades of Education Training Inzegane, Agadir, Morocco

Giorgio S. Senesi — CNR - Institute for Plasma Science and Technology (ISTP), Bari Seat, Bari, Italy

Hassan Nachit — Geoheritage and Geomaterials Laboratory, University Ibn Zohr, Agadir, Morocco

José Garcia — Petrographic Laboratory in Canary Museum of Meteorites, Tanta, Las Palmas, Canary Islands, Spain

Lahcen Ouknine — Geoheritage and Geomaterials Laboratory, University Museum of Meteorites, Ibn Zohr University, Agadir, Morocco

Mohamed Th. S. Heikal — Geology Department, Faculty of Science, Tanta University, Tanta, Egypt

INTRODUCTION

Meteorites are rocky or metalliferous fragments that have been ejected from a body of the solar system following impacts with other bodies and have arrived on Earth after traveling more or less long in space. The majority of meteorites come from the asteroid belt (rocky bodies orbiting between Mars and Jupiter) and some other meteorites, of lower frequency, can arrive from the Moon or the planet Mars. Meteorites provide information about the early stages of evolution of our solar system in general and Earth in particular, as they contain information about protosolar and even presolar material (DeMeo *et al.*, 2015; Gounelle, 2017).

After the great meteorite discoveries in Antarctica starting in 1976 (Corti *et al.* 2003), several arid regions of the world were found to be "deposits" of ancient meteorite falls (Bevan *et al.*, 1998). The hot deserts of the different continents are favorable places for their conservation, accumulation, and possible recovery. A large number of meteorites have been collected and studied in these places, namely: in Australia, the desert plain of Nullarbor (Bevan & Bindon, 1996; Bevan, 1998; Jull *et al.*, 2010), in North America, the desert region of Roosevelt County in the southwestern United States (Hutson *et al.*, 2013), in South America, the Atacama Desert in Chile (Hützler *et al.*, 2016; Valenzuela & Benado, 2018) and in Asia, the desert of Oman (Al-Kathiri *et al.*, 2005; Hofmann *et al.*, 2018), the Lut desert in Iran (Pourkhorsandi *et al.*, 2019) and the province of Xinjiang in China (Li *et al.*, 2017; Zeng *et al.*, 2018).

In Africa, where arid and hot zones represent 60% of the continent's surface, the first systematic studies were launched in the north of the continent (Sahara) at the end of the 20th century (Bischoff & Geiger, 1995; Schlüter *et al.* al., 2002; Ibhi, 2014, 2016; Khiri *et al.*, 2017; Ouknine *et al.*, 2019; Aboulahris *et al.*, 2019). In total, the number of meteorites recovered in Africa represents more than 1/6 of all extraterrestrial rocks around the world (Khiri *et al.*, 2017; Ouknine *et al.*, 2019), which makes this continent a region that shelters the second largest population of meteorites after that of Antarctica. This prompted us to consider the conditions/factors favorable to the collection of extraterrestrial rocks in Africa, both of meteorites collected after the observation of their fall (falls) and those found after a more or less long time of residence on earth (finds).

To do this we have subdivided our book into seven main chapters, briefly presented below.

The first chapter is devoted to a bibliographical synthesis of meteorites features that includes an exhaustive and concise description of the formation and origin of meteorites, their categorization into "falls" and "finds" according to the circumstances of their discovery, the criteria for their identification, the guidelines adopted for the nomenclature of meteorites falls and finds, and their classification.

The second chapter deals with the statistical study of falls recorded in various African countries by analyzing their spatiotemporal evolution and masses distribution, and by characterizing their typological classification in order to evaluate the different factors likely contributing to their observation and recovery.

The third chapter is devoted to the study of meteorite finds in Africa, their spatiotemporal distribution, and the human and natural factors that favor their discovery. Then, the distribution of meteorite masses collected and their typological classification are described. Subsequently, the variation of the degree of weathering (W) of African meteorites finds is analyzed, and the degree of influence of certain weathering factors, including climate, sample mass, earthly age and porosity, is evaluated.

The fourth chapter deals with North West Africa (NWA) meteorites in an attempt to contextualize and document these meteorites, define the circumstances of finding each sample, and assign the country of collection to each of them.

The fifth chapter is devoted specifically to Moroccan meteorites. The Moroccan territory is a privileged place for the collection of meteorites and is one of the richest countries in the world in terms of meteorites finds. The rate of recovery of meteorites (falls + finds) in Morocco exceeds that of most other countries of similar size and climatic conditions. More than 95% of documented meteorites from Morocco, including many rare types, have been recovered from Eastern Morocco Sahara, which has proved to be one of the most prolific areas in the world for meteorite finds.

The sixth chapter deals with Sahara meteorites. It is estimated that more than 90% of the surface of this desert has not yet been explored and that it still contains important meteorite falls. The most optimistic forecasts suggest that many new meteorites will continue to be recovered from the great Sahara Desert in the coming decades. Preserving the desert properly is essential to assure science such important research materials as meteorites.

The seventh chapter provides an up-to-date review of the confirmed and/or proposed meteorite falls and finds and their impact structures in Egypt, Sudan, and Libya. Among the ~190 confirmed impact structures/sites on Earth crust, only less than 8 have been identified in NE Africa, in particular in Oasis (Libya) and Kamil (Egypt) areas.

The book ends with some general conclusions highlighting the results obtained from the various studies performed on meteorites in Africa.

REFERENCES

Aboulahris, M., Chennaoui Aoudjehane, H., Rochette, P., Gattacceca, J., Jull, A.J.T., Laridhi Ouazaa, N., Folco, L., Buhl, S. (2019). Characteristics of the Sahara as a meteorite recovery surface. *Meteorit. Planet. Sci,* 5412, 2908-2928.
[http://dx.doi.org/10.1111/maps.13398]

Al-Kathiri, A., Hofmann, B.A., Jull, A.J.T., Gnos, E. (2005). Weathering of meteorites from Oman: Correlation of chemical and mineralogical weathering proxies with 14 C terrestrial ages and the influence of soil chemistry. *Meteorit. Planet. Sci,* 408, 1215-1239.
[http://dx.doi.org/10.1111/j.1945-5100.2005.tb00185.x]

Bevan, A.W.R., Bindon, P. (1996). Australian Aborigines and meteorites. *Rec. West. Aust. Mus,* 18, 93-102.

Bevan, A.W.R., Bland, P.A., Jull, A.L.T. (1998). Meteorite flux on the Nullarbor Region, Australia. *Meteorites: Flux with Time and Impact Effects. Special Publications 140,* London, U.K Geological Society 59-73.

Bischoff, A., Geiger, T. (1995). Meteorites from the Sahara: Find locations, shock classification, degree of weathering and pairing. *Meteorit. Planet. Sci,* 301, 113-122.

Corti, G., Zeoli, A., Bonini, M. (2003). Ice-flow dynamics and meteorite collection in Antarctica Earth & Planetary Science Letters, 215, pp. 371–378.

DeMeo, F.E., Alexander, C.M.O.D., Walsh, K.J., Chapman, C.R., Binzel, R.P. (2015). The compositional structure of the asteroid belt. *Asteroids IV,* 1, 13. [http://dx.doi.org/10.2458/azu_uapress_9780816532131-ch002]

Gounelle, M. (2017). *Les météorites : « Que sais-je? » n° 3859,* Presses universitaires de France [http://dx.doi.org/10.3917/puf.goune.2017.01]

Hutson, M., Ruzicka, A., Timothy Jull, A.J., Smaller, J.E., Brown, R. (2013). Stones from Mohave County, Arizona: Multiple falls in the "Franconia strewn field". *Meteorit. Planet. Sci,* 483, 365-389. [http://dx.doi.org/10.1111/maps.12062]

Hofmann, B.A., Gnos, E., Jull, A.J.T., Szidat, S., Majoub, A., Al Wagdani, K., Habibullah, S.N., Halawani, M., Hakeem, M., Al Shanti, M., Al Solami, A. (2018). Meteorite reconnaissance in Saudi Arabia. *Meteorit. Planet. Sci,* 5311, 2372-2394. [http://dx.doi.org/10.1111/maps.13132]

Hützler, A., Gattacceca, J., Rochette, P., Braucher, R., Carro, B., Christensen, E.J., Cournede, C., Gounelle, M., Laridhi Ouazaa, N., Martinez, R., Valenzuela, M., Warner, M., Bourles, D. (2016). Description of a very dense meteorite collection area in western Atacama: Insight into the long-term composition of the meteorite flux to Earth. *Meteorit. Planet. Sci,* 513, 468-482. [http://dx.doi.org/10.1111/maps.12607]

Ibhi, A. (2014). Morocco meteorite falls and finds: some statistics. International Letters of Chemistry. *Physics and Astronomy,* 1, 18-25.

Jull, A.J.T., McHargue, L.R., Bland, P.A., Greenwood, R.C., Bevan, A.W.R., Kim, K.J., LaMotta, S.E., Johnson, J.A. (2010). Terrestrial ages of meteorites from the Nullarbor region, Australia, based on 14C and 14C-10Be measurements. *Meteorit. Planet. Sci,* 458, 1271-1283. [http://dx.doi.org/10.1111/j.1945-5100.2010.01289.x]

Khiri, F., Ibhi, A., Saint-Gerant, T., Medjkane, M., Ouknine, L. (2017). Meteorite falls in Africa. *J. Afr. Earth Sci,* 134, 644-657. [http://dx.doi.org/10.1016/j.jafrearsci.2017.07.022]

Li, S., Wang, S., Leya, I., Smith, T., Tang, J., Wang, P., Zeng, X.J., Li, Y. (2017). A chondrite strewn field was found in east of Lop Nor, Xinjiang. *Kexue Tongbao,* 6221, 2407-2415. [http://dx.doi.org/10.1360/N972016-01450]

Ouknine, L., Khiri, F., Ibhi, A., Heikal, M.T.S., Saint-Gerant, T., Medjkane, M. (2019). Insight into African meteorite finds: Typology, mass distribution and weathering process. *J. Afr. Earth Sci,* 158, 103551. [http://dx.doi.org/10.1016/j.jafrearsci.2019.103551]

Pourkhorsandi, H., Gattacceca, J., Rochette, P., D'Orazio, M., Kamali, H., Avillez, R., Letichevsky, S., Djamali, M., Mirnejad, H., Debaille, V., Jull, A.J.T. (2019). Meteorites from the Lut Desert (Iran). *Meteorit. Planet. Sci,* 548, 1737-1763. [http://dx.doi.org/10.1111/maps.13311]

Schlüter, J., Schultz, L., Thiedig, F., Al-Mahdi, B.O., Aghreb, A.E.A. (2002). The Dar al Gani meteorite field (Libyan Sahara): Geological setting, pairing of meteorites, and recovery density. *Meteorit. Planet. Sci,* 378, 1079-1093. [http://dx.doi.org/10.1111/j.1945-5100.2002.tb00879.x]

Valenzuela, M., Benado, J. (2018). Meteorites and craters found in chile: a bridge to introduce the first attempt for geoheritage legal protection in the country. *Geoethics in Latin America,* ChamSpringer 103-115.
[http://dx.doi.org/10.1007/978-3-319-75373-7_7]

Zeng, X., Li, S., Leya, I., Wang, S., Smith, T., Li, Y., Wang, P. (2018). The Kumtag 016 L5 strewn field, Xinjiang Province, China. *Meteorit. Planet. Sci,* 536, 1113-1130.
[http://dx.doi.org/10.1111/maps.13073]

CHAPTER 1

Basic Issues on Meteorites: Origin, Formation, Identification, Nomenclature

Lahcen Ouknine[1,*], Fouad Khiri[1,2] and Abderrahmane Ibhi[1]

[1] *Geoheritage and Geomaterials Laboratory, Ibn Zohr University, Agadir, Morocco*

[2] *University Museum of Meteorites, Ibn Zohr University, Agadir, Morocco*

Abstract: Meteorites are rocky or metalliferous fragments that have been ejected from a body of the solar system following impacts with other bodies, and arriving on Earth after traveling more or less long in space. The majority of meteorites come from the asteroid belt (rocky bodies orbiting between Mars and Jupiter), and some other meteorites of lower frequency, can arrive from the Moon or the planet Mars. Meteorites provide information about the early stages of the evolution of our solar system in general and Earth in particular, as they contain information about protosolar and even presolar material. In this article, we will present an exhaustive synthesis of the formation and origin of meteorites, their categorization into "falls" and "finds" according to the circumstances of their discovery, the criteria for their identification, and the guidelines adopted for the nomenclature of meteorites falls and finds, and their classification.

Keywords: Categorization, Classification, Falls and finds meteorites, Identification, Nomenclature, Origin.

INTRODUCTION

Meteorites are fragments that were ejected from a body in the solar system following impacts with other bodies and arrive on Earth after a more or less long journey in space (Hughes, 1996). Rubin and Grossman (2010) offer a complete definition: a meteorite is a natural, solid object, larger than 10 μm, derived from a celestial body, which has been transported by natural means from the body on which it formed towards a region outside the dominant gravitational influence of this body, and which subsequently collided with a body larger than itself. Most of the known meteorites come from the asteroid belt (rocky bodies orbiting between Mars and Jupiter), and some other meteorites, of lower frequency, can arrive from the Moon or the planet Mars (Weisberg, 2018).

[*] **Corresponding author Lahcen Ouknine:** Geoheritage and Geomaterials Laboratory, Ibn Zohr University, Agadir, Morocco; E-mail: lahcen.ouknine@edu.uiz.ac.ma

Meteorites provide information about the early phases of evolution of our solar system in general and Earth in particular, as they contain information about protosolar, and even presolar, materials (DeMeo *et al.*, 2015; Gounelle, 2017).

ORIGIN OF METEORITES: VESTIGES OF THE EVOLUTION OF THE SOLAR SYSTEM

The Composition of the Solar System

The solar system formed 4.56 billion years ago from a cloud of gas and dust, the Protosolar Nebula. In a molecular cloud, the gravitational collapse of a dense, cold-core gives rise to a dense, hot protostar at its center (Aikawa *et al.*, 2008; Bardin, 2015) (Fig. **1**).

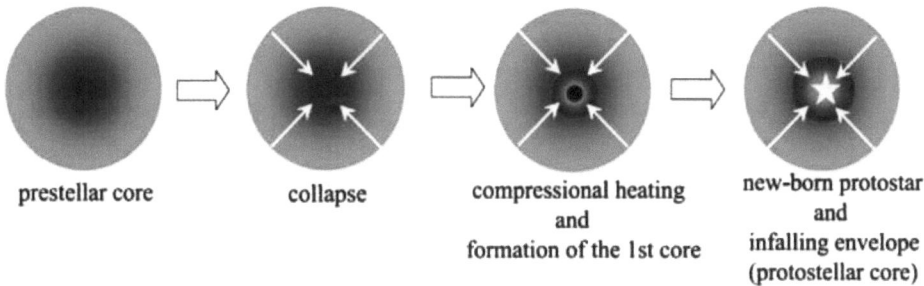

| prestellar core | collapse | compressional heating and formation of the 1st core | new-born protostar and infalling envelope (protostellar core) |

Fig. (1). Process of star formation, the Sun, for example (Aikawa *et al.*, 2008).

Some objects that did not accrete enough matter remained as small bodies, smaller than a few hundred km, such as asteroids and comets. Indeed, the solar system is composed of a star, 8 planets, 167 satellites, and a multitude of small objects such as dwarf planets (Pluto, Sedna, *etc.*), asteroids (Ceres, Vesta, Hebe, *etc.*), trans-Neptunian objects, and comets (Kuiper belt, Oort cloud). The solar system can be divided into four different regions:

The sun: a star that concentrates about 99.9% of the mass of the solar system and is composed mainly of hydrogen and helium and a few percent of the heavier elements. With a radius of 700,000 km, the sun is a member of the yellow dwarf star family and derives all of its energy through nuclear fusion (Chapman, 2007). Temperatures in the center reach around $10\text{-}15\times10^6$ °C and 6500 °C at the surface.

The inner solar system represents about 10% of the mass of the remaining solar system. It groups the 4 telluric planets, Mercury, Venus, Earth and Mars and their satellites, as well as the objects in the asteroid belt. Telluric planets are rocky and metallic bodies (rich in heavy elements), a few thousand km in diameter, with a thermal history, allowing mantle-core segregation (Bardin, 2015). In addition, the asteroid belt contains asteroids of small sizes (only about fifteen with a radius greater than 100 km), but of a very wide variety. However, we can find more or less evolved or differentiated bodies, rich in silicates, metal and even organic compounds or ice (Trigo-Rodríguez and Blum, 2008).

Outer solar system groups the 4 gas giant planets, Jupiter, Saturn, Uranus, and Neptune. These planets make up 90% of the remaining mass and are mostly made up of hydrogen and helium. These planets are made up of a gigantic gas envelope, and at their center, there could be a rocky and/or icy core. In addition, they present many ice and rocky satellites.

Distant solar system groups objects from the Kuiper belt, such as Pluton or Sedna, and the Oort cloud. Most of the bodies in this region are mainly composed of ice and dust. The comets would come from this zone, in particular from the Kuiper belt.

Formation and Ages of Meteorites

During the formation of the solar system, the first solids agglomerated to form planetesimals (Remusat, 2005). Like all bodies in the solar system, meteorites began to form in the early nebula together with the sun and the planets within the Milky Way, our galaxy. As already mentioned above, a nebula of gas and dust collapsed, contracted, and gave birth to the sun. The dust grains agglomerated to give larger and larger grains, resulting in the embryos of planets. This was the phenomenon of accretion (Aikawa *et al.*, 2008), which was accompanied by a very sharp increase in temperature. During this period, thermonuclear fusion appeared in the sun. In the rest of the cloud that divided into rings around the sun, the temperature started to drop, allowing the gas to condense into solid, future constituents of planetesimals and planets (Feigelson and Montmerle, 1999). After accretion, the original material of the Earth, planets and some asteroids (the largest) melted under a high temperature, and the mixture of homogeneous initial composition separated into several phases of different chemical compositions (DeMeo & Carry, 2014). In the case of planets, the starting material of solar composition was fractionated and separated into distinct layers: core (metallic), mantle (rich in olivine) and crust (silicate rocks).

Some objects that did not accrete enough matter remained as bodies smaller than a few hundred km, such as asteroids and comets. These are mostly undifferentiated objects that preserve the memory of the first million of years of the formation of the solar system (Bardin, 2015).

Origin of Meteorites

Meteorites are rocks detached from a parent body after impact with another celestial object (Weisberg *et al.*, 2006). The parent bodies of meteorites are either asteroids (Vesta, Ceres, *etc.*) or planets (Mars, Mercury, *etc.*) or our satellite, the Moon (Fig. **2**). As already noted, most meteorites originate from asteroids that orbit between Mars and Jupiter, in an area called the asteroid belt (Fig. **2**), which consists of billions of asteroids of different dimensions, ranging from a few cm to hundreds of km in diameter (Miguel, 2006; Trigo-Rodriguez and Blum, 2008). Asteroids are likely fragments of larger bodies that were disrupted by a collision with other bodies in the asteroid belt (Michel *et al.*, 2020).

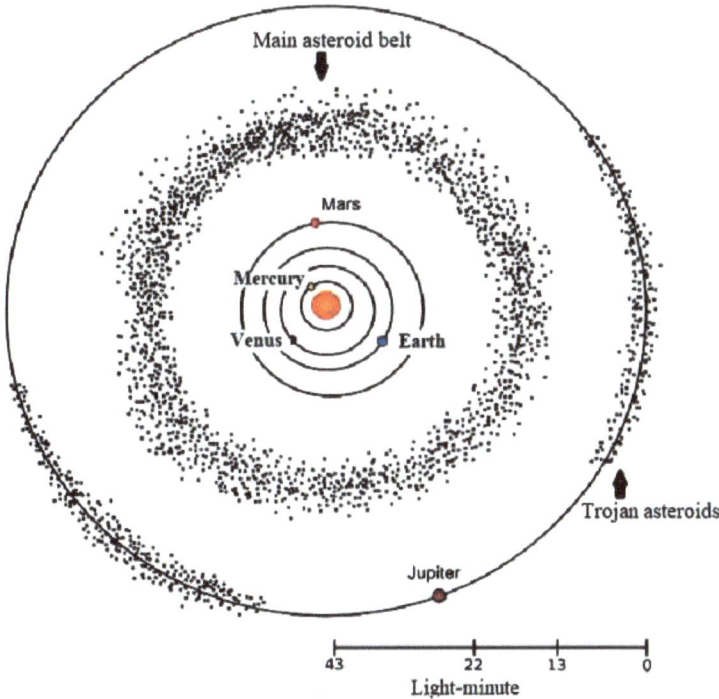

Fig. (2). Diagram of the inner solar system, showing the main asteroid belt. (Taken from: NASA website, https://www.nasa.gov/).

Historically, the British astronomer Greg, in 1854, was the first to propose that asteroids are the parent bodies of meteorites. Afterwards, research began, and appropriate evidence was provided to validate the hypothesis that asteroids would be shattered by collisions between them and their fragments be ejected out of their orbit. In addition, other asteroids that are not included in the asteroid belt can strike Mars or the Moon, and tear off pieces that can roam for a long time in space and finally fall on Earth as Martian or lunar meteorites, if their passage through space intersects with the Earth's orbit. In particular, they are slowed down by the Earth's atmosphere and the rapid compression of the air in front of the supersonic meteoroid results in heat transfer from the high-temperature gas to the surface of the meteoroid leading to a fusion of the outermost part of the meteorite (Weisberg *et al.*, 2006; DeMeo & Carry, 2014). Indeed, this crossing of the atmosphere changes the exterior appearance of the meteorite, which loses part of its material due to evaporation, melting and fragmentation and becomes covered with a black film (thickness <1 mm) called "fusion crust". The passage through the atmosphere also causes characteristic hollows, analogous to fingerprints, called regmaglyptes produced by the ablation of material (Taylor *et al.*, 2012).

Ages of Meteorites

Isotopic dating is the measurement of time using the decay of radioactive isotopes and accumulation of decay products at a known rate; isotopic chronometers can determine the time of the processes that fractionate parent and daughter elements. Modern isotopic dating can resolve time intervals of ~1 million years over the entire lifespan of the Earth and the Solar System. Using isotopic dating, a unified scale of time for the evolution of Earth, the Moon, Mars, and asteroids, and other planets can be built. Modern geochronology and cosmochronology rely on isotopic dating methods that are based on the decay of very long-lived radionuclides, *e.g.*, ^{238}U, ^{235}U, ^{40}K, *etc.*, to stable radionuclides, *e.g.*, ^{206}Pb, ^{40}K, ^{40}Ca, ^{87}Sr, ^{143}Rd, and moderately long-lived radionuclides, *e.g.*, ^{26}Al, ^{53}Mn, ^{146}Sm, ^{182}Hf, and stable nuclides, *e.g.*, ^{26}Mg, ^{53}Cr, ^{142}Nd, ^{182}W. The diversity of physical and chemical properties of parent (radioactive) and daughter (radiogenic) nuclides, their geochemical and cosmochemical affinities, and the resulting diversity of processes that fractionate parent and daughter elements allow the direct isotopic dating of a vast range of terrestrial and planetary processes (Amelin, 2020).

In particular, radiochronological studies show that meteorites date between 4.3 to 4.6 billion years ago (Podosek, 1972; Connelly *et al.*, 2012). At least three different ages can be measured for a meteorite (Fig. 3).

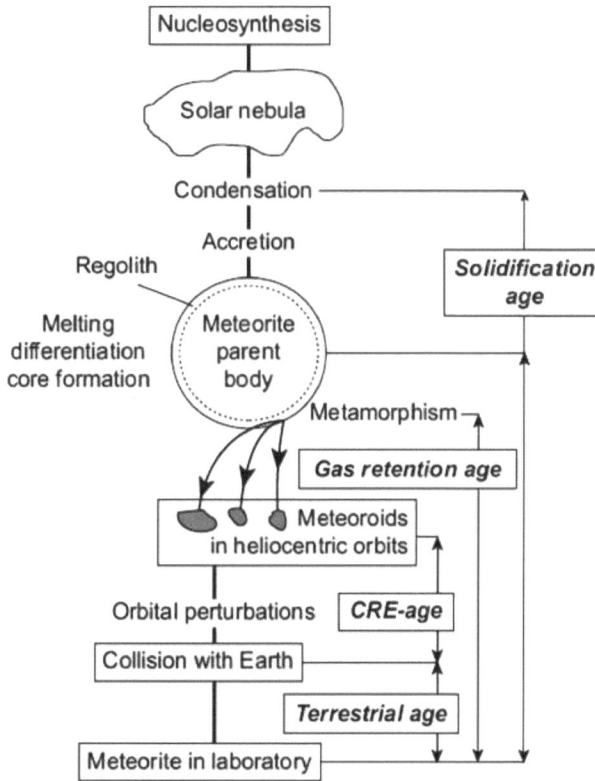

Fig. (3). Schematic representation of the different ages of a meteorite (Lipschutz & Schultz, 2014).

The absolute age measures the age of formation of the original rock by solidification of the initial liquid, the last chemical differentiation, and the cooling of meteoritic materials (Wasson, 2012). This age is about 4.56 billion years for most meteorites with the exception of some much younger Martian and lunar meteorites that feature more recent geological activity. The dating of meteorites is indeed measured by isotopic analyses of very long-lived chemical elements (for example, dating by ^{87}Rb-^{87}Sr).

The age of exposure to cosmic radiation measures the time interval between the moment the meteorite is detached from its parent body, probably following a shock and its capture and fall on Earth. During this trip, the rock fragments are exposed for the first time to cosmic radiation. When a particle of cosmic radiation strikes a fragment, it causes nuclear reactions leading to the formation of new atoms. The crystals that make up the future meteorite record these reactions in the form of traces that can be detected quite easily. In particular, trace density provides a measure of the time the fragment spent in space after being detached, as it was previously protected by the layers that covered it (Gispert, 2010). Then, the time elapsed from its ejection from the parent asteroid until its arrival on Earth, *i.e.*, "the exposure time", can be determined often by measuring ^{3}He, ^{21}Ne, and ^{38}Ar. For example, the age of exposure of the Tissint meteorite was measured to be 0.7 ± 0.3 Myears (Chennaoui *et al.*, 2012; Ibhi *et al.*, 2013).

The age of residence or Earth age defines the residence time of a meteorite on the Earth surface, when its fall cannot be observed. Unfortunately, meteorites are quite vulnerable during this time as they are subject to erosion and oxidation as soon as they reach the ground. In general, the dating of the terrestrial age of meteorites is performed by short-lived radioactive isotopes such as ^{14}C (T1 / 2 = 5730 years), ^{36}Cl (T1 / 2 = 0.3 × 10^6 years), ^{26}Al (T1 / 2 = 0.72 × 10^6 years). For recently fallen (a few years) meteorites, the age of residence is measured directly by gamma radiation of cosmogenic isotopes, such as ^{22}Na (2.6 years) and ^{54}Mn (312 days).

CATEGORIZATION AND NOMENCLATURE OF METEORITES

Categorization of Meteorites: Falls and Finds

Only meteoroids (often resulting from the partial disintegration of an asteroid) of a size generally greater than several tens of cm will survive passage through the atmosphere, whereas smaller objects disintegrate completely and do not reach the ground (Koschny *et al.*, 2019). Indeed, each year from 2000 to 5000 meteorites that weigh more than a kg fall on our planet, but 75% of them disappear into the oceans, whereas only a few of the 25% remaining ones can be occasionally recovered on the dependence on meteorology and nature of the ground (Halliday *et al.*, 1989). The Meteorite Nomenclature Committee (NomCom) of the Meteoritical Society, the only body authorized to receive official meteorite declarations in the world, has divided meteorites into two categories according to the circumstances of their discovery (Agee, 2015a).

Falls or observed falls, if at least part of the event leading to the fall of a meteorite on Earth, has been observed either by humans (by the naked eye) or by devices (cameras, radar, *etc.*). For a meteorite to be declared a fall, two conditions must be assessed, *i.e.*, the fall event must be well documented, and the recovered meteorite must be related to the fall event. Generally, the pieces of the meteoroid fragmented in the atmosphere distribute within an "ellipse of fall" (Trigo-Rodríguez, 2006), which allows one to roughly define the geographical extent of the fall resulting from the trajectory followed by the meteoroid entering the Earth's atmosphere and producing an intense luminous phenomenon together with its fragmentation. As the smallest fragments generally travel less far than the large ones, the distribution of the fragments according to their mass allows us to estimate the direction of the bolide and can help to determine its orbit, which will provide sufficient indications on the place of fall. Statistical studies and direct observations made by automatic cameras indicate a rate of around 500 bright bolides entering the Earth's atmosphere each year, of which only ten are recovered in the form of meteorites (Zolensky *et al.*, 2006).

Finds, if the meteorite fall is not observed, and it is recovered after spending time on Earth (Agee, 2015a). The finds represent most meteorites collected on Earth (98%), especially in Antarctica. Since 1969, more than 47000 meteorites have been recovered from the blue ice in Antarctica by scientific expeditions (https://www.lpi.usra.edu/meteor/metbull.php, January 2023). The large number of meteorites in this region is favored by an accumulation mechanism, *i.e.*, the meteorites which fall on the Antarctic ground are then buried more and more deeply due to the accumulation of snow, then they are transported with the ice as it flows to the edge of the continent. If a relief obstructs this flow, the ice tends to rise on the relief, so meteorites accumulate on the surface of the ice. One region in Antarctica is famous for its accumulations of meteorites, *i.e.*, the Allan Hills region, from which the name ALH (followed by a number) was given to meteorites collected in this area. Furthermore, over the past twenty years, meteorite collection has increased exponentially to finds in the hot deserts of Morocco, Algeria, Libya and Oman, the North American and Australian deserts, and Atacama in Chile.

In December 2014, the Meteorite Nomenclature Committee (NomCom) of the Meteoritical Society adopted a new categorization system for falls and finds (Agee, 2015b). Instead of the binary system Fall-Find, the new system consists of five categories, *i.e.*, Confirmed fall, Probable fall, Find-possible fall, Find-doubtful fall, and Find. Currently, 78479 meteorites are listed in the Meteoritical Society database (Meteoritical Bulletin), of which only 1370 are falls, and all remaining ones are finds (https://www.lpi.usra.edu/meteor/metbull.php, January 2023). Once officially classified, a meteorite is given a unique name that generally refers to the place of collection.

Nomenclature of Meteorites

According to the new meteorite nomenclature guide adopted by the nomenclature committee, unique names are given to all individual meteorites, whether they are falls or finds (Agee *et al.*, 2015b). Furthermore, the name of a meteorite must be clearly distinct from any other meteorite name and abbreviation, and must indicate the geographic location of the fall or find. However, in areas where meteorite finds occur in large numbers, the numbering and listing systems are different. In particular, in dense collection areas (DCA) of meteorites, which are located mainly in deserts, numbered find sequences can amount to thousands of samples, as in the case of Dhofar 001 to Dhofar 1982 in Oman and that of Dar al Gani 001 to Dar al Gani 1069 in the Libyan desert. Differently, falling meteorites are given unique place names, even when occurring within the confines of a DCA. For example, a fall observed in 1998 in the Roosevelt County DCA in New Mexico, USA, was called Portales Valley. Similar names, such as Wethersfield (1971) and Wethersfield (1982), are attributed to two rare meteorite falls in the same location (Gattacceca *et al.*, 2018). Finally, if the provenance of a new meteorite is unknown or disputed, it gets the name "Nova" and a digital index, *i.e.*, a sequential number, as in the case of several samples found in Northwest Africa, *i.e.*, NWA 001 to NWA 10969.

In general, the classification and nomenclature of new meteorites (falls and finds) by the Nomenclature Committee requires a wealth of information such as: a) the location of the fall or find (geographic coordinates); b) the circumstances of the fall or discovery; c) the total known mass and the number of parts recovered; d) the notified typology of the meteorite; e) the location of the main mass, *etc.* (Agee *et al.*, 2015b).

CRITERIA FOR THE IDENTIFICATION OF METEORITES

Macroscopic Recognition

In the field, observation with the naked eye makes it possible to identify samples likely to be meteorites and to rule out any rocks which are likely to be of terrestrial origin. The most important macroscopic criteria are:

Fusion crust which consists of a glassy film, usually black in color, of molten rock covering the exterior of a freshly fallen meteorite (Fig. **4a**). Usually, the fusion crust is black due to the presence of oxidized iron on the meteorite surface formed during its crossing through the atmosphere.

Widmanstätten figures which generally appear on the surface of metallic meteorites after polishing and etching with a mixture of nitric acid and alcohol (Nital), and consisting of intergrowths of kamacite and taenite (Fig. **5b**). The Widmanstätten texture can be attributed to the solid-state transition of low-Ni taenite to aggregate of high Ni taenite and kamacite crystals.

Regmaglyptes, or "thumb marks" which are frequent but sometimes not easily identifiable indentations present on the surface of the meteorite (Fig. **4b**), and consist of molten material torn from the surface of the meteorite during its passage through the atmosphere.

Chondrules and metals which consist of mm-sized rounded objects and white or yellow metallic particles (iron-nickel metal or sulfides, respectively) (Fig. **5a**) that appear on a fresh break or a polished surface of a meteorite. The chondrules are sometimes visible in the fusion crust.

(a) (b)

Fig. (4). Fusion crust (Benguerir, LL6) **(a)** and regmaglypts (NWA 2932, L6) **(b)** (Credit: University Museum of Meteorites, © Ibhi A.).

(a) (b)

Fig. (5). A mega-chondrule on an ordinary chondrite (NWA 869, L3-6) **(a)** and Widmanstätten structures on a polished section of an iron meteorite (Taza Meteorite, NWA 859) **(b)** (Credit: University Museum of Meteorites, © Ibhi A.).

Density which is generally higher for meteorites than terrestrial rocks. Carbonaceous chondrites have a density between 2.2 and 2.9, while metallic meteorites have a density between 7.5 and 7.9 (Heide and Wlotzka, 1995).

Vanishing lines which consist of lines located at the level of the fusion crust testifying to the orientation of the meteorite upon arrival on Earth (oriented meteorite) (Fig. **6a**). They appear like drops of fluid being pushed very energetically backward by high velocity.

Magnet attraction which is featured by chondrites and other metal-bearing meteorites and is due to the high presence of Fe-Ni metal phase, whereas achondrite meteorites show a very weak magnetization (Fig. **6b**).

(a)   (b)

Fig. (6). An oriented meteorite (ordinary chondrite) **(a)** (© wiki.meteoritica.pl.) and the magnet reaction of an ordinary chondrite (H5) **(b)** (Credit: University Museum of Meteorites, © Ibhi A.).

Magnetic susceptibility which is the rate of magnetization induced in a material subjected to a weak magnetic field (<1mT) (Rochette *et al.* 2009, Ostrowski & Bryson, 2019). This parameter can be measured by a fast, non-destructive technique that allows to determine the amount of metals or other magnetic minerals in a sample (Smith *et al.*, 2006) and the rapid classification of new meteorites according to the diagram in Fig. (**7**). However, this method cannot be used alone because the domains of some meteorites overlap.

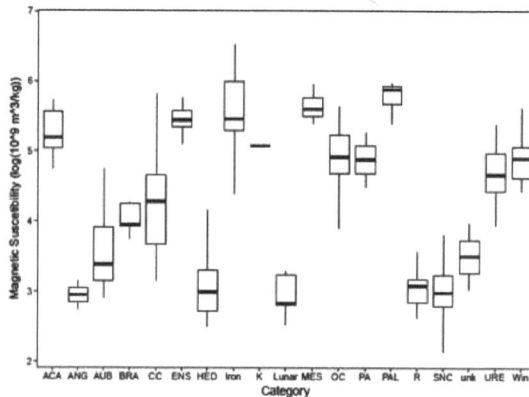

Fig. (7). Magnetic susceptibility diagram (logχ). Average logχ (in 10^{-9} m³/kg) of meteorites, including chondrite falls (Ostrowski & Bryson, 2019).

Microscopic Recognition

Recognizing a meteorite with the naked eye is often difficult due to the similarity of certain types (achondrites, for example) to magmatic rocks. Indeed, most minerals present in meteorites are also found in magmatic, basic and ultrabasic terrestrial rocks (Bland *et al.*, 2006). It is, therefore, necessary to consider several criteria and carry out mineralogical and geochemical analyses to accurately classify meteorites (Ibhi, 2016). Thus, the recognition of minerals often begins with the observation of the phases with a binocular magnifying glass, then with an optical microscope in transmitted light mode and by a more detailed observation using scanning electron microscopy (SEM) (Fig. **8**). In particular, SEM combined with energy dispersive X-ray spectroscopy (EDS) analysis can also provide the chemical composition of meteorites. Actually, together with optic microscopy, it covers >90% of tasks concerning the identification and classification of meteorites.

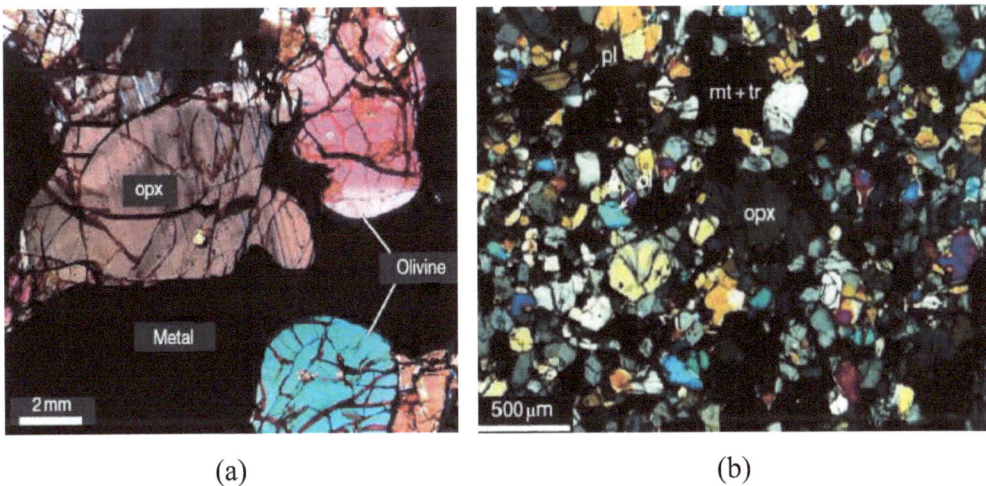

(a)   (b)

Fig. (8). Photomicrographs obtained by a transmission microscope of two thin sections of two meteorites: pallasite Milton (**a**) and acapulcoite ALH 81315 (**b**) (Mt: FeNi-metal, Opx: orthopyroxene, Ol: olivine, Pl: plagioclase and Tr: Troilite) (Krot *et al.*, 2014).

The measurement of oxygen isotopes is another advanced technique used to prove the extraterrestrial origin of meteorites, and is also used to confirm the classification based on chemical composition (Clayton *et al.*, 1976; Clayton, 1993). The graphical representation of $\delta17O$ versus $\delta18O$ shows that the split lines of different types of meteorites are distinct from those of terrestrial rocks, except for lunar meteorites that share a common reservoir of oxygen isotopes with Earth (Javoy, 1995, Clayton & Mayeda, 1999) (Fig. **9**).

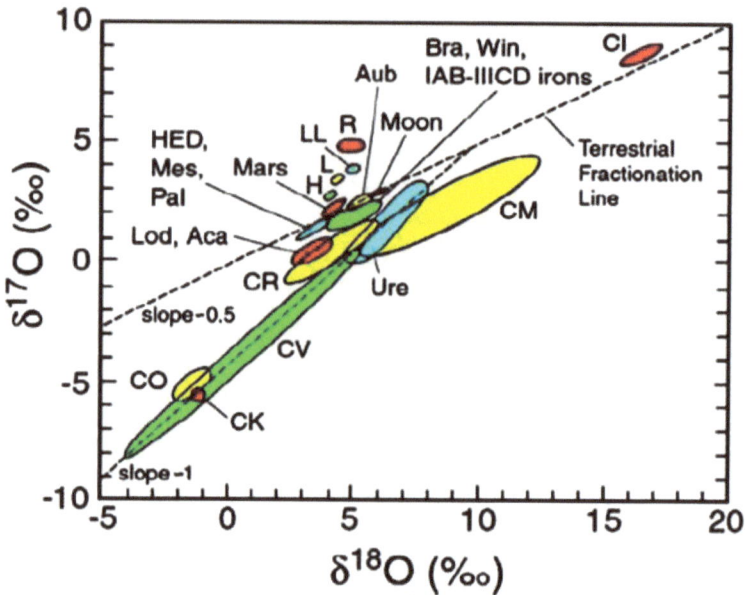

Fig. (9). Oxygen isotope diagram of split lines of different classes of meteorites (Clayton & Mayeda, 1999).

Furthermore, meteorites recovered from space missions have helped to decide on the typology and origin of certain meteorites (*e.g.*, lunar meteorites). In particular, four missions allowed the return of extraterrestrial samples: i) the Apollo (1969-1972) and Luna (1970-1976) missions allowed the return of lunar samples (in total 380 kg and 320 g, respectively); ii) the Stardust mission (launched in 1999) brought back to Earth 100 μg of dust from the comet Wild 2; iii) the Genesis probe (2001-2004) captured solar wind particles with the aim of better understanding the formation and functioning of Sun; iv) the Hayabusa probe (2005) returned to Earth in June 2010 with 1534 tiny samples of mineral dust from the asteroid Itokawa. Furthermore, other missions are in course in recent years, *i.e.*, Chang-e (2020) on the Moon, Hayabusa 2 (2014) on the Riugu, and Osiris-Rex (2016) on the near-Earth asteroid Bennu.

CLASSIFICATION OF METEORITES

Meteorites are classified to sort into very similar types and allow a better understanding of their origin and relationships. According to their composition, meteorites are divided into three large families (Tschermak, 1885; Rubin, 1997): stony meteorites (chondrites and achondrites), mixed meteorites (mesosiderites and pallasites), and iron meteorites (hexahedrites, octahedrites, and ataxites). Furthermore, considering their origin and formation process, meteorites are subdivided into undifferentiated chondrites, primitive achondrites, and differentiated achondrites (Weisberg *et al.*, 2006). In 2014, Krot *et al.* proposed a more detailed classification that prioritizes the different types of meteorites into classes, groups, and subgroups (Fig. **10**) based on primary criteria, such as petrography, mineralogy, geochemistry, relationships between chemical elements (*e.g.*, Fe and Mn), isotopic composition of oxygen *etc.*, and secondary criteria, such as degree of impairment and level of shock. As a result, meteorites sharing similar characteristics are thus grouped within the same group. For example, chondrites are divided into three broad classes - carbonaceous (C), ordinary (O) and enstatite (E). However, there are a few meteorites that are not affiliated with any group or subgroup, known as "ungrouped".

Fig. (10). Classification of meteorites (Krot *et al.*, 2014).

Undifferentiated Meteorites: Chondrites

Chondrites represent 85% of meteorite falls, are considered "fossils" of the early solar system, and originate from parent bodies with a diameter of less than a few tens of km that have not yet initiated the processes of differentiation. The constituent material of chondrites has agglomerated in the protoplanetary disc, and they are over 4.5 billion years old, formed in the solar nebula at a time when no planet yet existed around the sun (Beatty *et al.*, 1999). Chondrites take their name from the small spherules they contain, *i.e.*, chondrules, mainly made up of silicate minerals, mainly olivine and pyroxenes, and resulting from the crystallization of small droplets brought to a liquid state by fusion processes occurred in the solar nebula (Wood, 1990). They were probably formed during the first 1 to 3 million years of solar accretion by shock waves caused by gravitational instabilities (Desch *et al.*, 2012). Shortly after accretion, some chondrites experienced a heating up to temperature ranging from 200 to 1000°C that was insufficient to cause their fusion, thus this rise in temperature transformed the structure of chondrites and the composition of their minerals, but only slightly modified their overall chemical composition (Desch & Connolly, 2002). This metamorphism was more or less marked depending on whether the meteorite originated from the surface of the asteroid, where the temperature was low, or from its center. Thus, chondrites are classified according to a metamorphism scale of petrologic types ranging from 1 to 6. In particular, type 3 chondrites are the best preserved, and the most representative of the primitive nebula state; chondrites from types 3 to 1 result increasingly transformed by water (H_2O), and chondrites of types 3 to 6 are gradually metamorphosed by undergoing an increasingly strong temperature rise (Koschny *et al.*, 2019). The most visible effect of metamorphism is the recrystallization of the matrix and the progressive erasure of the chondrules. In general, chondrites are constituted (Fig. **11**) by: (i) the metal (metallic iron associated with nickel in variable amounts depending on the type of chondrite); (ii) the matrix (silicate dust containing olivine, pyroxene, amorphous silicates and a certain amount of metal); and (iii) refractory Ca-Al-rich inclusions (CAIs), which consist of diffuse zones containing refractory minerals, such as aluminous and titaniferous pyroxene, anorthite and spinel that are accessory components in some types of carbonaceous chondrites and are rarely found in other chondrite groups (Greshake & Fritz, 2018).

According to their chemical composition, the family of chondrites is divided into several classes: carbonaceous chondrites (CC), ordinary chondrites (CO), enstatite chondrites (EC), type R chondrites (rumurutites), and type K (Kakangari) chondrites.

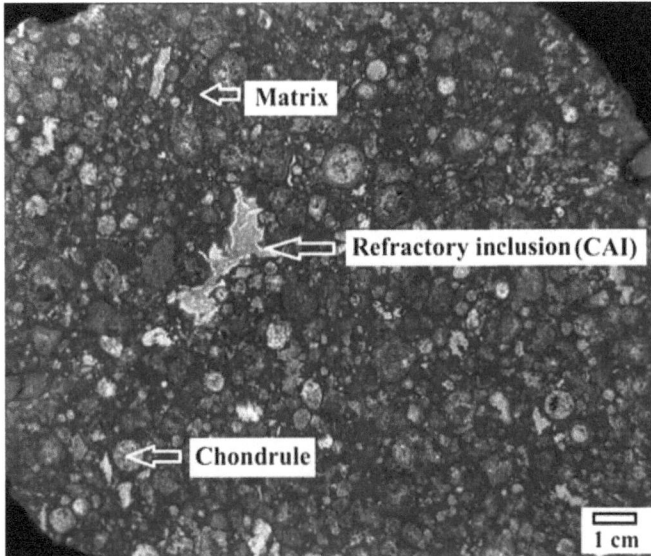

Fig. (11). Chondrules in a slice of a piece of the Allende carbonaceous chondrite (Greshake & Fritz, 2018).

Carbonaceous Chondrites

Carbonaceous chondrites (CC) are primitive and undifferentiated stony meteorites composed of chondrules bathed in a fine-grained silicate matrix in which CAIs are representative of the first material condensed from the hot nebula (Abreu & Brearley, 2010). The CC are also characterized by the presence of organic compounds, including long-chain hydrocarbons and amino acids similar to those occurring in protein synthesis. Carbonaceous meteorites have undergone significant aqueous alteration, and many of them have escaped thermal metamorphism (Trigo-Rodríguez, 2015). Depending on the degree of thermal metamorphism and aqueous alteration they have undergone on the parent bodies (C asteroids), carbonaceous chondrites are subdivided into various chemical groups, *i.e.*, CI, CM, CV, CO, CK, CR, CH, and CB. In particular, the CI and CM groups exhibit a significant abundance of water, mainly in the form of hydrated minerals, whereas the CO, CV and CR chondrites underwent much less severe water alteration.

CI group was named after the Ivuna meteorite that fell in Tanzania in 1938 (Kallemeyn & Wasson, 1981). They contain almost 26% water and a certain amount of organic matter (amino acids) and are devoid of chondrules (Krot, 2006).

CM group was named after the Mighei meteorite that fell in Ukraine in 1889 (Kallemeyn & Wasson, 1981). They contain 3 to 11% water and a few chondrules. Kallemeyn and Wasson (1981) found that these meteorites contain over 230 different amino acids, while only 20 are present on the Earth.

CV group was named by analogy to the Vigarano meteorite that fell in Italy in 1910 (Kallemeyn & Wasson, 1981). These meteorites are characterized by their richness in refractory CAIs and large chondrules (5 to 25 mm) (Krot *et al.*, 1995).

CO group, named after the Ornans meteorite that fell in France in 1868 (Kallemeyn & Wasson, 1981). These meteorites show a lower presence of chondrules compared to those of the CV group and contain less than 1% water and smaller CAIs sparsely distributed in the matrix (Kallemeyn & Wasson, 1981).

CK group, named after the Karoonda meteorite that fell in Australia in 1930 (Kallemeyn & Wasson, 1981). This group shows a degree of metamorphism greater than that of other groups and features a large quantity of magnetite dispersed in a matrix rich in olivine and pyroxene.

CR group, named after the Renazzo meteorite that fell in Italy in 1981 (Kallemeyn & Wasson, 1981). These meteorites are characterized by the abundance of iron-nickel in both metallic (5 to 8%) and sulphide (1 to 4%) forms, with respect to the other groups (Kallemeyn & Wasson, 1981).

CH group, named for their richness in iron (H: high Iron) and their no relation to a specific locality. They contain Fe, Ni and very little C, and present very small chondrules.

CB group, named after the Benccubin meteorite that fell in Australia in 1930. These meteorites are mainly characterized by the richness in metal in the form of spheres (0.1 to 10 mm) welded with chondritic silicates (Krot *et al.*, 2004).

Ordinary Chondrites

Ordinary chondrites (OC) constitute more than 85% of meteorite falls worldwide. They are characterized by a variable ratio of silicates and metal and a great abundance of large spheroidal chondrules (of mm sizes) of different textures and mineral compositions (Weisberg *et al.*, 2006). This class is divided into three groups: H, L and LL, which can be recognized on a chemical basis, in particular by their amount of iron compared to silicon (Norton, 2002). Group H features a high iron content by weight (27%), group L contains 23% iron, and group LL 20% iron. Actually, ordinary chondrites are assigned to distinct groups on the basis of fayalite (Fa) content of olivine and ferrosilite (Fs) content of pyroxenes (Kallemeyn *et al.*, 1989) (Fig. **12**).

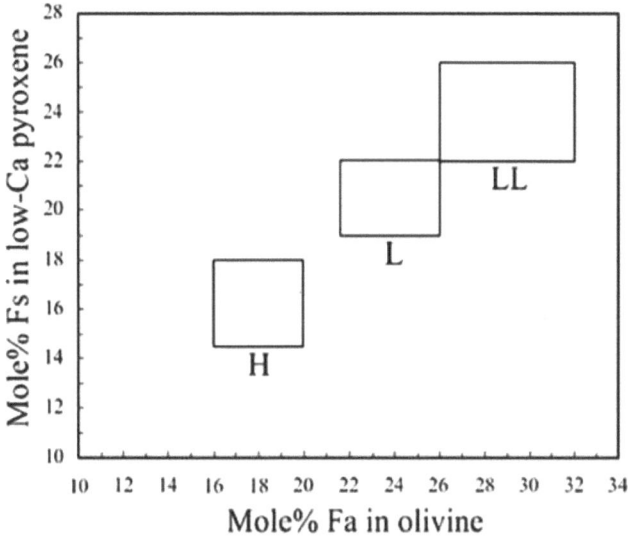

Fig. (12). Classification of ordinary chondrites according to the fayalite contents of olivine and ferrosilite contents of pyroxenes (Kallemeyn *et al.*, 1989).

Ordinary chondrites are formed in separate parent asteroids that exhibit different petrographic types (3 to 6), reflecting the depth and rate of cooling. The study of nearly 50 000 classified OC-type meteorites has shown that they have undergone different degrees of thermal metamorphism (Moyano-Cambero *et al.*, 2017). Thus, the more the recrystallization of the matrix and the chondrules, the higher the petrographic type. The degree of brecciation increases as the petrographic type decreases (closer to the surface). For ordinary chondrites, the process of thermal metamorphism requires a parent body diameter between 160 and 180 km (Sears *et al.*, 1991).

Enstatite Chondrites

Enstatite chondrites (EC) are characterized by iron in metal or sulphide forms, such as troilite (FeS), and silicates, such as low iron pyroxene of the enstatite type. Enstatite chondrites are subdivided into two groups, EH (high enstatite) and EL (low enstatite), which are differentiated by their iron and sulfur content. In particular, the EH group contains small chondrules (0.2 mm) and 30% total iron, while the EL group contains larger chondrules (> 0.5 mm) and 25% total iron (Keil, 1989). Each group includes thermally metamorphosed chondrite types from 3 to 6.

R-type Chondrites

The designation of this group is based on only one fall recorded, *i.e.*, Rumuruti, that fell in Kenya in 1934. These chondrites are highly oxidized, rich in olivine but poor in iron. They differ from other chondritic groups by their oxidation state, oxygen isotope composition and mineralogy (Weisberg *et al.*, 2006).

K-type Chondrites

The chondrites in this group are named after their type specimen Kakangari, a meteorite that fell in Tamil Nadu, India, in 1890. They represent one of the rarest meteorite groups in the world. All K chondrites known to date belong to the petrographic type 3. They are rich in ferrous sulfide and troilite and feature many chondrules. The K chondrites are unique in their chemical composition and exhibit an isotopic oxygen signature that distinguishes them from all other chondritic groups. These features suggest that K chondrites must have originated in a small, primitive parent body that has not yet been identified (Weisberg *et al.*, 2006).

Achondrites

Achondrites are differentiated meteorites that represent only 7.8% of meteorites collected globally. They originate from large parent bodies with diameters of ten to several hundred km that made it possible to initiate differentiation that resulted in the reorganization of matter within the initial chondritic bodies. Thanks to the partial fusion, gravitational segregation occurred between a dense core mainly composed of the heaviest liquids (iron-nickel) and a less dense mantle and rocky crust consisting of the lightest liquids (silicates). Achondrites are divided into three classes, each representing one of the various parts of the parent body, *i.e.*, differentiated achondrites, silicate achondrites (originating from the crust and/or the mantle of the parent body), ferrous (originating from the core of the parent body) and "mixed meteorites" (pallasites and mesosiderites) that originate in part from the interface between the metallic core and the stony mantle, although other hypotheses have been proposed (Tarduno *et al.*, 2012). Another group of achondrites, *i.e.*, "primitive achondrites", which feature remains of chondrules but have undergone a partial fusion, are considered either as a separate class at the same rank as chondrites and achondrites (Weisberg *et al.*, 2006; Krot *et al.*, 2014), or as a subclass belonging to achondrites (Bischoff *et al.*, 2001). This subclass of meteorites confirms the continuity between differentiated and undifferentiated meteorites.

Primitive Undifferentiated Achondrites

Primitive undifferentiated achondrites are similar to chondrites and have almost the same age (Krot *et al.*, 2003). Some samples show distinct chondrules that prove that this group represents the transition between chondrites and achondrites (Greenwood *et al.*, 2007). This class of meteorites is subdivided into three main groups.

Lodranites, named after the Lodran type specimen, a meteorite that fell in Pakistan in 1868. Lodranites are a good example of early achondrites with chondritic chemistry and achondritic texture (McCoy *et al.*, 1997). They present a matrix with coarse crystals without any residual chondrule, which is the result of advanced metamorphism (T between 1050 and 1200°C) (McCoy *et al.*, 1997).

Acapulcoites, named after the Acapulco meteorite that fell in Mexico in 1976. They are mainly composed of olivine, orthopyroxene, plagioclase, metallic iron-nickel, and iron sulphide (troilite). Acapulcoites resemble lodranites but exhibit a certain difference related to the degree of thermal metamorphism (T between 950 and 1000°C), which reveals chondrules different from lodranites (McCoy *et al.*, 1997).

Winonaites, named after the Winona meteorite that was found in a prehistoric site (Elden) in Arizona, USA, in 1928. Winonaites are mainly composed of large crystals of pyroxene, olivine, troilite and metallic iron-nickel (the total iron content varies between 18% and 30%) (Benedix *et al.*, 1998).

Differentiated Achondrites

Differentiated Asteroidal Achondrites

They originate from parent bodies (asteroids) with diameters of several hundred km. Under the effect of heating caused by the decay of radioactive isotopes, these objects melted, and the heaviest elements formed metallic cores (Ni and Fe), while the lighter elements (silicates) formed a mantle and a rocky crust (Weisberg *et al.*, 2006; Weisberg, 2018). The class of differentiated achondrites is divided into three groups.

Brachinites, named after the Brachina type meteorite discovered in Australia. This group of meteorites are classified either as primitive achondrites or as asteroidal achondrites. They are chondritic in composition, but exhibit achondritic textures (Mittlefehldt, 2003). Brachinites are distinguished from other asteroidal meteorites by their very high olivine content that varies between 74 to 98% by volume (Nehru *et al.*, 1996).

Ureilites, sometimes classified as primitive achondrites due to their age of formation that is around 4.5 billion years. Ureilites are mainly composed of olivine and secondarily pigeonite, troilite and iron–nickel, and contain a matrix rich in graphite and nanodiamonds (high carbon content, on average 3% by weight) (Norton, 2002). Nanodiamonds are probably the result of high-pressure shock waves produced by the collision of the parent body with other asteroids (Goodrich *et al.*, 1987).

Howardites, Eucrites and Diogenites (HEDs), which are metal-poor achondrites originated from the asteroid Vesta and featured an iron-nickel alloy segregated into the metal core and a silicate mantle and crust that is the source of HED (Ruzicka *et al.*, 1997). This group of achondrites is subdivided into several subgroups according to their mineralogical composition (Warren, 1985).

Howardites, which are brecciated achondrites forming a mixture of basaltic rock (Eucrites) and ultrabasic cumula (Diogenites) with a diogenetic fraction of less than 9%. These meteorites originated by one or more violent impacts that detached a significant thickness of the asteroid crust, thus freeing the materials necessary for meteorite formation by re-agglomeration (surface eucritical and very deep diogenetic).

Eucrites, which originated from the cooling of the magma near the surface of their parent body, and are subdivided into cumulative eucrites, non-cumulative eucrites and polymictic eucrites.

Diogenites are achondrites composed mainly of orthopyroxene and accessory minerals such as olivine, plagioclase, and spinel (Mittlefehldt, 1994).

Aubrites

Aubrites are predominantly composed of magnesian orthopyroxene, like enstatite chondrites. The isotopic composition of oxygen ranges between that of aubrites and enstatite chondrites, which would indicate the same zone of formation in the early solar system (Clayton & Mayeda, 1984).

Angrites

These achondrites have a granular texture and are composed of three main minerals, *i.e.*, clinopyroxene, olivine and plagioclase. The origin of these meteorites is controversial between an asteroidal origin (Mittlefehldt, 2003) and a planetary origin (Mercury) (Markowski *et al.*, 2007).

Planetary Achondrites

Planetary achondrites are differentiated meteorites originating from the Moon or the planet Mars. They were torn from their parents bodies by very violent impacts.

Martian achondrites, also named SNC after three meteorites that fell in three distinct regions: Shergottite (after the Shergotty region in India), Nakhlite (after Nakhla in Egypt) and Chassignite (after Chassigny in France). Recently, a fourth type has been added to the group of Martian meteorites, *i.e.*, Orthopyroxenite, which is a single sample (ALH 8400, Antarctica) (Mittlefehldt, 1994). *Shergottites* represent the vast majority of Martian meteorites, and consist of basalts torn from the surface of Mars near volcanic terrain (McSween, 2002). Other Martian meteorites belong to the group of: (i) *Nakhlites* which are composed mainly of clinopyroxene and formed from basaltic magma around 1.3 billion years ago; (ii) *Chassignites*, which contain more than 95% olivine (dunites) and incidentally pyroxenes and chromiferous spinel (McSween, 2002); and (iii) *Orthopyroxenite* that is composed by over 98% orthopyroxene and accessory minerals such as plagioclase and certain carbonates (Mittlefehldt, 1994).

Lunar achondrites, which show a very wide chemical and mineralogical variety similar to those brought back by Apollo and Luna missions (Papike *et al.*, 1998). According to their aluminum content (Al_2O_3) and the ratio between ferrous oxide and magnesium oxide (FeO/MgO), Le Bas (2001) subdivided the lunar meteorites recovered on the Earth into four groups: (i) *Basalts*, also called "basalt ponds", which are basalt-type materials from the Lunar Seas featuring a low content of titanium oxide compared to those brought back by Apollo (Basilevsky *et al.*, 2010); (ii) *Anorthosites*, which are monomictic brecciated meteorites rich in calcium and aluminum (Le Bas, 2001), mainly composed of calco-sodium plagioclase feldspar, and originated from the lunar "highlands" (continental highlands); (iii) *Mixed breccias*, which originate from lunar surfaces between the basalt seas and the continental high plateaus, are composed of various rock fragments and are subdivided into regolithic breccias rich or poor in siderophilic elements (Korotev, 1999); and (iv) *Gabbros*, which are plutonic rocks very high in olivine originated from the fusion of primitive mantle materials of peridotite type (such as terrestrial gabbros).

Iron Meteorites

The group of iron meteorites represents about 5.7% of known meteorites, and is composed of more than 90% iron and nickel, and the rest consists of nodules of silicates, sulphides and graphite. These meteorites are mainly characterized by

Widmanstätten structures visible on sections previously treated with nitric acid. The mechanism of formation of iron meteorites consisted in the segregation of iron into silicates due to heating with the decay of certain isotopes such as ^{26}Al (Haack & McCoy, 2003). Then, the ferronickel melt crystallized into taenite (rich in nickel, 20% to 65%) and kamacite (poor in nickel, 4 to 7.5%) (Chabot & Drake, 1999) which resulted in a spherical body featuring a metallic core and a rocky crust.

Depending on the presence and distribution of Widmanstätten structures, iron meteorites are subdivided into three subgroups (Haack & McCoy, 2003): (i) *Ataxites*, which are rich in nickel (between 16 and 60%), the crystals are very small and the Widmanstätten figures are visible only by a microscope; (ii) *Octahedrites* which are the most numerous and are characterized by a nickel content from 7 to 15%) and by the presence of clearly visible Widmanstätten structures; and (iii) *Hexahedrites*, which are formed mainly by kamacite, feature a low nickel content (between 5 and 6%), and do not present Widmanstätten figures, but often show thin parallel lines called "Neumann lines" that were generated by the shocks undergone by hexahedrites.

Another chemical classification system was proposed to define the chemical subgroups of iron meteorites, which is based on the contents of nickel and trace elements such as gallium, germanium, iridium, platinum and gold (Scott & Wasson, 1975; Wasson *et al.*, 1989). Specific diagrams based on this approach separate iron meteorites into 13 types, *i.e.*, IAB, IC, IIAB, IIC, IID, IIE, IIF, IIG, IIICD, IIIE, IIIF, IVA and IVB.

Mixed or Stony-iron Meteorites

Mixed meteorites are intermediate between iron and stony meteorites and contain as many silicates as metals. They are quite rare, representing no more than 1% of known meteorites, and are subdivided into two sub-groups.

Mesosiderites, which are composed of metamorphosed plagioclases and pyroxenes, similar to those in HED achondrites, encompass metal fragments (Haack & McCoy, 2003).

Pallasites, which are composed of magnesium-rich olivine embedded in a metallic matrix that originated from a core/mantle mixture of a type IIIAB parent body where slow crystallization allowed olivine phenocrysts to develop and be embedded in a solidifying metallic liquid (Yang *et al.*, 2010).

CONCLUSION

Meteorites are fascinating rocks that arrive on Earth from space. They provide unique information useful to understand the evolution of the solar system, and, thus, the formation of the Earth. However, meteorites formed under conditions very different from those on the Earth today, thus are not chemically stable and change during their permanence on the Earth. Meteorites can be collected after their fall or searched systematically in regions where the conditions for accumulation and preservation are favorable. In the next articles of this book, more details are provided on meteorite falls and finds in Africa. This vast continent with immense semi-arid to arid zones is believed to be a reservoir of a very significant number of meteorites.

REFERENCES

Abreu, N.M., Brearley, A.J. (2010). Early solar system processes recorded in the matrices of two highly pristine CR3 carbonaceous chondrites, MET 00426 and QUE 99177. *Geochim. Cosmochim. Acta, 74*(3), 1146-1171.
[http://dx.doi.org/10.1016/j.gca.2009.11.009]

Agee, C., Bullock, E., Bouvier, A., Dunn, T., Gattacceca, J., Grossman, J., Herd, C., Ireland, T., Metzler, K., Mikouchi, T., Ruzicka, A., Smith, C., Welten, K., Welzenbach, L. (2015). Categorization of Finds and Falls. *Meteoritical Society.* Available at: http://www.lpi.usra.edu/meteor/docs /falls-finds.pdf(2015a).

Agee, C., Bullock, E., Bouvier, A., Dunn, T., Gattacceca, J., Grossman, J., Herd, C., Ireland, T., Metzler, K., Mikouchi, T., Ruzicka, A., Smith, C., Welten, K., Welzenbach, L. Guidelines for meteorite nomenclature. Available at: http://meteoriticalsociety.org/?page_id=59(Accessed on: 2015b).

Amelin, Y. (2020). Isotopic Dating.*Oxford Research Encyclopedia of Planetary Science..* Oxford University Press.
[http://dx.doi.org/10.1093/acrefore/9780190647926.013.133]

Aoudjehane, H.C., Avice, G., Barrat, J.A., Boudouma, O., Chen, G., Duke, M.J.M., Franchi, I.A., Gattacceca, J., Grady, M.M., Greenwood, R.C., Herd, C.D.K., Hewins, R., Jambon, A., Marty, B., Rochette, P., Smith, C.L., Sautter, V., Verchovsky, A., Weber, P., Zanda, B. (2012). Tissint martian meteorite: a fresh look at the interior, surface, and atmosphere of Mars. *Science, 338*(6108), 785-788.
[http://dx.doi.org/10.1126/science.1224514] [PMID: 23065902]

Aikawa, Y., Wakelam, V., Garrod, R.T., Herbst, E. (2008). Molecular evolution and star formation: from prestellar cores to protostellar cores. *Astrophys. J., 674*(2), 984-996.
[http://dx.doi.org/10.1086/524096]

Bardin, N. (2015). Composition isotopique des éléments légers dans les micrométéorites ultracarbonées par spectrométrie de masse à émission ionique secondaire à haute résolution en masse, contribution à la connaissance des surfaces cométaires.*Thèse de doctorat* Université Paris-Saclay.

Bas, M.J.L. (2001). Report of the Working Party on the classification of the lunar igneous rocks. *Meteorit. Planet. Sci., 36*(9), 1183-1188.
[http://dx.doi.org/10.1111/j.1945-5100.2001.tb01953.x]

Basilevsky, A.T., Neukum, G., Nyquist, L. (2010). The spatial and temporal distribution of lunar mare basalts as deduced from analysis of data for lunar meteorites. *Planet. Space Sci., 58*(14-15), 1900-1905.
[http://dx.doi.org/10.1016/j.pss.2010.08.020]

Beatty, K.J., Beatty, J.K., Petersen, C.C., Chaikin, A. (1999). *The new solar system..* Cambridge University Press.

Benedix, G.K., McCoy, T.J., Keil, K., Bogard, D.D., Garrison, D.H. (1998). A petrologic and isotopic study of winonaites: evidence for early partial melting, brecciation, and metamorphism. *Geochim. Cosmochim. Acta, 62*(14), 2535-2553.
[http://dx.doi.org/10.1016/S0016-7037(98)00166-5]

Bischoff, A. (2001). Meteorite classification and the definition of new chondrite classes as a result of successful meteorite search in hot and cold deserts. *Planetary and Space Science., 49*(8), 769-776.
[http://dx.doi.org/10.1016/S0032-0633(01)00026-5]

Bland, P.A., Zolensky, M.E., Benedix, G.K., Sephton, M.A. (2006). Weathering of chondritic meteorites. *Meteorites and the early solar system II., 853-867.
[http://dx.doi.org/10.2307/j.ctv1v7zdmm.45]

Chapman, C.R. (2007). The Asteroid Impact Hazard and Interdisciplinary Issues.*Comet/Asteroid Impacts and Human Society.*. Berlin, Heidelberg: Springer.
[http://dx.doi.org/10.1007/978-3-540-32711-0_7]

Clayton, R.N. (1993). Oxygen isotopes in meteorites. *Annu. Rev. Earth Planet. Sci., 21*(1), 115-149.
[http://dx.doi.org/10.1146/annurev.ea.21.050193.000555]

Clayton, R.N., Mayeda, T.K. (1999). Oxygen isotope studies of carbonaceous chondrites. *Geochim. Cosmochim. Acta, 63*(13-14), 2089-2104.
[http://dx.doi.org/10.1016/S0016-7037(99)00090-3]

Clayton, R.N., Mayeda, T.K., Rubin, A.E. (1984). Oxygen isotopic compositions of enstatite chondrites and aubrites. *J. Geophys. Res., 89*(S01), C245.
[http://dx.doi.org/10.1029/JB089iS01p0C245]

Clayton, R.N., Onuma, N., Mayeda, T.K. (1976). A classification of meteorites based on oxygen isotopes. *Earth Planet. Sci. Lett., 30*(1), 10-18.
[http://dx.doi.org/10.1016/0012-821X(76)90003-0]

Connelly, J.N., Bizzarro, M., Krot, A.N., Nordlund, Å., Wielandt, D., Ivanova, M.A. (2012). The absolute chronology and thermal processing of solids in the solar protoplanetary disk. *Science, 338*(6107), 651-655.
[http://dx.doi.org/10.1126/science.1226919] [PMID: 23118187]

DeMeo, F.E., Carry, B. (2014). Solar System evolution from compositional mapping of the asteroid belt. *Nature, 505*(7485), 629-634.
[http://dx.doi.org/10.1038/nature12908] [PMID: 24476886]

DeMeo, F.E., Alexander, C.M.O.D., Walsh, K.J., Chapman, C.R., Binzel, R.P. (2015). The compositional structure of the asteroid belt. *Asteroids IV, 1*, 13.
[http://dx.doi.org/10.2458/azu_uapress_9780816532131-ch002]

Desch, S.J., Connolly, H.C., Jr (2002). A model of the thermal processing of particles in solar nebula shocks: Application to the cooling rates of chondrules. *Meteorit. Planet. Sci., 37*(2), 183-207.
[http://dx.doi.org/10.1111/j.1945-5100.2002.tb01104.x]

Desch, S.J., Morris, M.A., Connolly, H.C., Jr, Boss, A.P. (2012). The importance of experiments: Constraints on chondrule formation models. *Meteorit. Planet. Sci., 47*(7), 1139-1156.
[http://dx.doi.org/10.1111/j.1945-5100.2012.01357.x]

Feigelson, E.D., Montmerle, T. (1999). High-energy processes in young stellar objects. *Annu. Rev. Astron. Astrophys., 37*(1), 363-408.
[http://dx.doi.org/10.1146/annurev.astro.37.1.363]

Gispert, J. (2010). *Astronomie générale.* Association Marseillaise d'Astronomie.

Goodrich, C.A., Jones, J.H., Berkley, J.L. (1987). Origin and evolution of the ureilite parent magmas: Multi-stage igneous activity on a large parent body. *Geochim. Cosmochim. Acta, 51*(9), 2255-2273.
[http://dx.doi.org/10.1016/0016-7037(87)90279-1]

Gounelle, M. (2017). *Les météorites: « Que sais-je? » n° 3859.*. Presses universitaires.de France:

[http://dx.doi.org/10.3917/puf.goune.2017.01]

Greenwood, R.C., Franchi, I.A., Gibson, J.M., Benedix, G.K. (2007). Oxygen isotope composition of the primitive Achondrites. *Lunar & Planetary Science Conference, 38*, 2163.

Greshake, A., Fritz, J. (2018). Meteorites.*Planetary Geology.* (pp. 103-121). Cham: Springer. [http://dx.doi.org/10.1007/978-3-319-65179-8_6]

Haack, H., McCoy, T.J. (2003). Iron and stony-iron meteorites.*In: Treatise on Geochemistry* Elsevier.(pp. 325-346). Oxford.:

Halliday, I., Blackwell, A.T., Griffin, A.A. (1989). The flux of meteorites on the Earth's surface. *Meteoritics, 24*(3), 173-178. [http://dx.doi.org/10.1111/j.1945-5100.1989.tb00959.x]

Heide, F., Wlotzka, F. (1995). The Meteorites.Meteorites. Berlin, Heidelberg: Springer. 97–168. [http://dx.doi.org/10.1007/978-3-642-57786-4_3]

Hughes, D.W. (1996). The size, mass and evolution of the Solar System dust cloud. *Quarterly Journal of the Royal Astronomical Society., 37*, 593.

Ibhi, A. (2013). New Mars meteorite fall in Morocco: final strewn field. *Physics and Astronomy, 11*, 20-25. [http://dx.doi.org/10.18052/www.scipress.com/ILCPA.16.20]

Ibhi, A. (2013). Meteors and meteorite falls in Morocco. International Letters of Chemistry. *Physics and Astronomy, 12*, 28-35.

Ibhi, A. (2016). Météorites : Perles du Désert Marocain. ISBN. 978-9954-37-760-4. 244.

Javoy, M. (1995). The integral enstatite chondrite model of the Earth. *Geophys. Res. Lett., 22*(16), 2219-2222. [http://dx.doi.org/10.1029/95GL02015]

Kallemeyn, G.W., Rubin, A.E., Wang, D., Wasson, J.T. (1989). Ordinary chondrites: Bulk compositions, classification, lithophile-element fractionations and composition-petrographic type relationships. *Geochim. Cosmochim. Acta, 53*(10), 2747-2767. [http://dx.doi.org/10.1016/0016-7037(89)90146-4]

Kallemeyn, G.W., Wasson, J.T. (1981). The compositional classification of chondrites—I. The carbonaceous chondrite groups. *Geochim. Cosmochim. Acta, 45*(7), 1217-1230. [http://dx.doi.org/10.1016/0016-7037(81)90145-9]

Keil, K. (1989). Enstatite meteorites and their parent bodies. *Meteoritics, 24*(4), 195-208. [http://dx.doi.org/10.1111/j.1945-5100.1989.tb00694.x]

Korotev, R.L. (1999). A new estimate of the composition of the feldspathic upper crust of the Moon.*In Lunar & Planetary Science Conference, 30*, 1303. [http://dx.doi.org/10.1111/j.1945-5100.1989.tb00694.x]

Koschny, D., Soja, R.H., Engrand, C. (2019). Space Science Reviews Springer. 215, p.34.

Krot, A.N., Keil, K., Goodrich, C.A., Scott, E.R.D., Weisberg, M.K. (2003). Classification of meteorites. In meteorites, comets, and planets. *Treatise on Geochemistry, 1*, 83-128.

Krot, A.N., Keil, K., Scott, E.R.D., Goodrich, C.A., Weisberg, M.K. (2014). Classification of meteorites and their genetic relationships. 1, pp. 83–128. [http://dx.doi.org/10.1016/B978-0-08-095975-7.00102-9]

Krot, A.N., Scott, E.R.D., Zolensky, M.E. (1995). Mineralogical and chemical modification of components in CV3 chondrites: Nebular or asteroidal processing? *Meteoritics, 30*(6), 748-775. [http://dx.doi.org/10.1111/j.1945-5100.1995.tb01173.x]

Krot, A.N., Amelin, Y., Russell, S.S., Twelker, E. (2004). Are chondrules in the CB carbonaceous chondrite Gujba primary (nebular) or secondary (asteroidal)? *Meteorit. Planet. Sci., 39*, 56.

Markowski, A., Quitté, G., Kleine, T., Halliday, A., Bizzarro, M., Irving, A. (2007). Hafnium–tungsten chronometry of angrites and the earliest evolution of planetary objects. *Earth Planet. Sci. Lett., 262*(1-2), 214-229.
[http://dx.doi.org/10.1016/j.epsl.2007.07.035]

McCoy, T.J., Keil, K., Muenow, D.W., Wilson, L. (1997). Partial melting and melt migration in the acapulcoite-lodranite parent body. *Geochim. Cosmochim. Acta, 61*(3), 639-650.
[http://dx.doi.org/10.1016/S0016-7037(96)00365-1]

McSween, H.Y., Jr (2002). The rocks of Mars, from far and near. *Meteorit. Planet. Sci., 37*(1), 7-25.
[http://dx.doi.org/10.1111/j.1945-5100.2002.tb00793.x]

Meteoritical bulletin database *The Meteoritical Society.* Available at: http://www.lpi.usra.edu/meteor(2014).

Michel, P., Ballouz, R.L., Barnouin, O.S., Jutzi, M., Walsh, K.J., May, B.H., Manzoni, C., Richardson, D.C., Schwartz, S.R., Sugita, S., Watanabe, S., Miyamoto, H., Hirabayashi, M., Bottke, W.F., Connolly, H.C., Yoshikawa, M., Lauretta, D.S. (2020). Collisional formation of top-shaped asteroids and implications for the origins of Ryugu and Bennu. *Nat. Commun., 11*(1), 2655.
[http://dx.doi.org/10.1038/s41467-020-16433-z] [PMID: 32461569]

Miguel, D. (2006). *Les météorites, ces cailloux tombés du ciel.* FUNDP.

Mittlefehldt, D.W. (1994). The genesis of diogenites and HED parent body petrogenesis. *Geochim. Cosmochim. Acta, 58*(5), 1537-1552.
[http://dx.doi.org/10.1016/0016-7037(94)90555-X]

Mittlefehldt, D.W., Lindstrom, M.M. (2003). Geochemistry of eucrites: genesis of basaltic eucrites, and Hf and Ta as petrogenetic indicators for altered antarctic eucrites. *Geochim. Cosmochim. Acta, 67*(10), 1911-1934.
[http://dx.doi.org/10.1016/S0016-7037(02)01411-4]

Moyano-Cambero, C.E., Pellicer, E., Trigo-Rodríguez, J.M., Williams, I.P., Blum, J., Michel, P., Küppers, M., Martínez-Jiménez, M., Lloro, I., Sort, J. (2017). Nanoindenting the Chelyabinsk meteorite to learn about impact deflection effects in asteroids. *Astrophys. J., 835*(2), 157.
[http://dx.doi.org/10.3847/1538-4357/835/2/157]

Nehru, C.E., Prinz, M., Weisberg, M.K., Ebihara, M.E., Clayton, R.N., Mayeda, T.K. (1996). A new brachinite and petrogenesis of the group. *Lunar & Planetary Science Conference, 27*, 943-944.

Norton, O.R. (2002). Book Review: The Cambridge encyclopedia of meteorites/Cambridge University Press. *Observatory, 122*, 297p.

Ostrowski, D., Bryson, K. (2019). The physical properties of meteorites. *Planet. Space Sci., 165*, 148-178.
[http://dx.doi.org/10.1016/j.pss.2018.11.003]

Papike, J.J. (1998). *Lunar samples.* Planetary materials.

Podosek, F.A. (1972). Gas retention chronology of Petersburg and other meteorites. *Geochim. Cosmochim. Acta, 36*(7), 755-772.
[http://dx.doi.org/10.1016/0016-7037(72)90086-5]

Remusat, L.M. (2005). Étude moléculaire et isotopique en deutérium de la matière organique insoluble des chondrites carbonées. *Thèse de Doctorat. Muséum national d'histoire naturelle., 190.*

Rochette, P., Gattacceca, J., Bourot-Denise, M., Consolmagno, G., Folco, L., Kohout, T., Pesonen, L., Sagnotti, L. (2009). Magnetic classification of stony meteorites: 3. Achondrites. *Meteorit. Planet. Sci., 44*(3), 405-427.
[http://dx.doi.org/10.1111/j.1945-5100.2009.tb00741.x]

Rubin, A.E. (1997). A History of the Mesosiderite Asteroid: Enigmatic meteorites, mixtures of iron and stone, can be understood by unraveling the evolutionary history of their parent asteroid. *Am. Sci., 85*(1), 26-35.

Rubin, A.E., Grossman, J.N. (2010). Meteorite and meteoroid: New comprehensive definitions. *Meteorit. Planet. Sci., 45*(1), 114-122.

Ruzicka, A., Snyder, G.A., Taylor, L.A. (1997). Vesta as the howardite, eucrite and diogenite parent body: Implications for the size of a core and for large-scale differentiation. *Meteorit. Planet. Sci., 32*(6), 825-840.
[http://dx.doi.org/10.1111/j.1945-5100.1997.tb01573.x]

Scott, E.R.D., Wasson, J.T. (1975). Classification and properties of iron meteorites. *Rev. Geophys., 13*(4), 527-546.
[http://dx.doi.org/10.1029/RG013i004p00527]

Sears, D.W., Batchelor, J.D., Lu, J., Keck, B.D. (1991). Metamorphism of CO and CO-like chondrites and comparisons with type 3 ordinary chondrites. *Antarctic Meteorite Research.*

Smith, D.L., Ernst, R.E., Samson, C., Herd, R. (2006). Stony meteorite characterization by non-destructive measurement of magnetic properties. *Meteorit. Planet. Sci., 41*(3), 355-373.
[http://dx.doi.org/10.1111/j.1945-5100.2006.tb00468.x]

Tarduno, J.A., Cottrell, R.D., Nimmo, F., Hopkins, J., Voronov, J., Erickson, A., Blackman, E., Scott, E.R.D., McKinley, R., McKinley, R.D. (2012). Evidence for a dynamo in the main group pallasite parent body. *Science, 338*(6109), 939-942.
[http://dx.doi.org/10.1126/science.1223932] [PMID: 23161997]

Taylor, S., Matrajt, G., Guan, Y. (2012). Fine-grained precursors dominate the micrometeorite flux. *Meteorit. Planet. Sci., 47*(4), 550-564.
[http://dx.doi.org/10.1111/j.1945-5100.2011.01292.x]

Trigo-Rodríguez, J.M., Blum, J. (2008). The role of collisional compaction in primitive asteroids and comets. *European Planetary Science Congress, 29.*

Tschermak, G. (1885). The microscopic properties of meteorites. Die Mikroskopische Beschaffenheit der Meteoriten. *Smithsonian Contributions to Astrophysics., 4*(6), 138-239.
[http://dx.doi.org/10.5479/si.00810231.4-6.138]

Warren, P.H. (1985). Origin of howardites, diogenites and eucrites: A mass balance constraint. *Geochim. Cosmochim. Acta, 49*(2), 577-586.
[http://dx.doi.org/10.1016/0016-7037(85)90049-3]

Wasson, J.T. (2012). *Meteorites: Classification and properties.* Springer Science and Business Media.

Wasson, J.T., Ouyang, X., Wang, J., Eric, J. (1989). Chemical classification of iron meteorites: XI. Multi-element studies of 38 new irons and the high abundance of ungrouped irons from Antarctica. *Geochim. Cosmochim. Acta, 53*(3), 735-744.
[http://dx.doi.org/10.1016/0016-7037(89)90016-1]

Weisberg, M.K. (2018). Meteorites. Encyclopedia of geochemistry encyclopedia of earth sciences series book series (EESS) springer. 917-924.

Wood, J.A. (1990). *Meteorites.* The New Solar System.

Yang, J., Goldstein, J.I., Scott, E.R.D. (2010). Main-group pallasites: Thermal history, relationship to IIIAB irons, and origin. *Geochim. Cosmochim. Acta, 74*(15), 4471-4492.
[http://dx.doi.org/10.1016/j.gca.2010.04.016]

Zolensky, M., Bland, P., Brown, P., Halliday, I. (2006). Meteorites and the early solar system II. University of arizona press. 869-888.
[http://dx.doi.org/10.2307/j.ctv1v7zdmm.46]

CHAPTER 2

Meteorite Falls in Africa

Fouad Khiri[1,2,3,*], **Abderrahmane Ibhi**[1,2] and **Lahcen Ouknine**[1,2]

[1] *Geoheritage and Geomaterials Laboratory, Ibn Zohr University, Agadir, Morocco*

[2] *University Museum of Meteorites, Ibn Zohr University, Agadir, Morocco*

[3] *Regional Center of Trades of Education and Training, Inzegane, Agadir, Morocco*

Abstract: Collecting meteorites just after their fall is a fundamental element to continue to gather information on the history of our solar system. During the period 1800-2020, 170 observed meteorite falls were recorded in Africa. The mass of fragments collected for any African meteorite range from 1.4 g to 175 kg, with a predominance of cases from 1 to 10 kg. The average rate of observed falls in Africa is low, with only one recovery per 1.29 years (*i.e.*, 0.026 per year and per million km^2). The African collection of observed falls is dominated by chondrites (84.4%), as in the world collection. The achondrites (10%) include three famous Martian meteorite falls: Nakhla (Egypt), Tissint (Morocco), and Zagami (Nigeria), whereas the observed iron meteorite falls are relatively rare (*i.e.*, 5% of the collection). The rate of documented falls in Africa has been increasing since 1860, with 88% recovered during the period 1910-2020. Most of these falls have been observed and then collected in North-Western Africa, Eastern Africa and Southern Africa, in countries that feature a large area and a large but evenly distributed population. Other factors that are proven to be favorable to the observation and collection of meteorite falls on the African territory are a genuine meteorite education, the semi-arid to arid climate offering clear skies most of the time, cultivated land or sparse grassland and the possible access to the fall location favored by a low percentage of forest cover and a dense road network.

Keywords: Africa, Classification, Distribution factors, Observed meteorite falls, Statistics.

INTRODUCTION

Meteorites have persisted unaltered since their extremely old formation (4.55 Ga) (Wasson & Wetherill, 1979), thus, when they reach the Earth, they represent an invaluable source of information not yet available about the solar system history, its planets, and the question of life origin if meteorites contain organic molecules (Gounelle, 2009). For these reasons, meteorite collection is important, particularly

* **Corresponding author Fouad Khiri:** Geoheritage and Geomaterials Laboratory, Ibn Zohr University, Agadir, Morocco, University Museum of Meteorites, Ibn Zohr University, Agadir, Morocco & Regional Center of Trades of Education and Training, Inzegane, Agadir, Morocco; E-mail: fkhiri2009@gmail.com

in the case of observed falls that offer fresh material for scientific research. The Halliday study (2001) based on photographic data of fireballs from Canadian network cameras collected over eleven years shows that 4500 events per year dropped at least 1 kg of meteorites on Earth's land surface. Unfortunately, several of these extraterrestrial fragments disappeared as shooting stars (Kress, 2001) were lost while diving in the oceans or remained unnoticed on the land. The annual average of meteorites recovered around the world over the past two centuries does not exceed five to six pieces (Graham *et al.*, 1985).

The main objective of this chapter is to describe, based on the analysis of statistical data, the "observed meteorite falls" in Africa in terms of number and class in time and space, and to examine the contribution of certain demographic and geographical factors to their recovery. Africa has been chosen due to the importance of its scientific contribution to the research and study of meteorites and the great scientific and cultural value of some meteorites falls in this continent. These include the carbonaceous meteorite Tarda (Meteoritical Bulletin Database, 2020) and the observed Martian falls Nakhla and Tissint (Ibhi, 2013a; Ibhi *et al.*, 2013; Treiman, 2003) that contain extraterrestrial organic molecules (Treiman, 1993; Beech, 2003; McCubbin *et al.*, 2013; Lin *et al.*, 2014). The choice is also justified by the large surface area of the African continent (20% of all terrestrial area, *i.e.,* 30 million km^2) (FAO, 2002), its geographical location between the northern and southern hemispheres and the diversity of its landscapes (desert, forests, mountains) and climates. A significant part of the meteorite falls on Earth is hosted in this large area.

TEMPORAL AND SPATIAL EVOLUTION AND DISTRIBUTION OF METEORITE FALLS IN AFRICA

Based on the information provided in the Meteoritical Bulletin Database (www.lpi.usra.edu/meteor/metbull.php), spatial and temporal data on observed meteorite falls in 52 African countries between 1801 and 2020 have been collected, retaining only those falls that have been approved by the Meteorite Nomenclature Committee of the Meteoritical Society, whereas geographical and demographic data were obtained from various FAO reports.

Evolution of Numbers, Masses, and Classes of African Observed Meteorite Falls

Evolution of Falls Numbers

Since 1800, the year when meteorites were recognized as objects falling from the sky, scientists have recorded 170 observed meteorite falls in Africa, totaling a mass of 2914.7 kg. The oldest meteorite fall (L6) dates back to 1801 in Mauritius, while the most recent one, dated August 20, 2020, is a fragment of more than 4 kg of a carbonaceous chondrite (C2-ung) that exploded in the Tarda region in northern Morocco (Weather bulletin database, 2020).

Over 12.6% of all known meteorite falls worldwide through December 31, 2020, have been recorded in Africa. The number of these falls is similar to that recorded in the USA (163) and is higher than that of other regions in similar time periods (in Russia, 54 observed falls have been inventoried since 1805). These results confirm the important contribution of the African continent to the total observed meteorite falls. Furthermore, the ratio of falls to finds in Africa (1:60) is lower compared to that recorded in other continents or large countries of similar size (Australia 1:41, South America 1:17; United States 1:13; Canada 1:6; Russia 1:3 and India 1:1).

The cumulative number of observed meteorite falls since 1800 in Africa shows a constant increase (Fig. **1**). The recovery rate for these falls averages 0.77 falls per year over the past 220 years and 0.026 falls per year per 106 km^2.

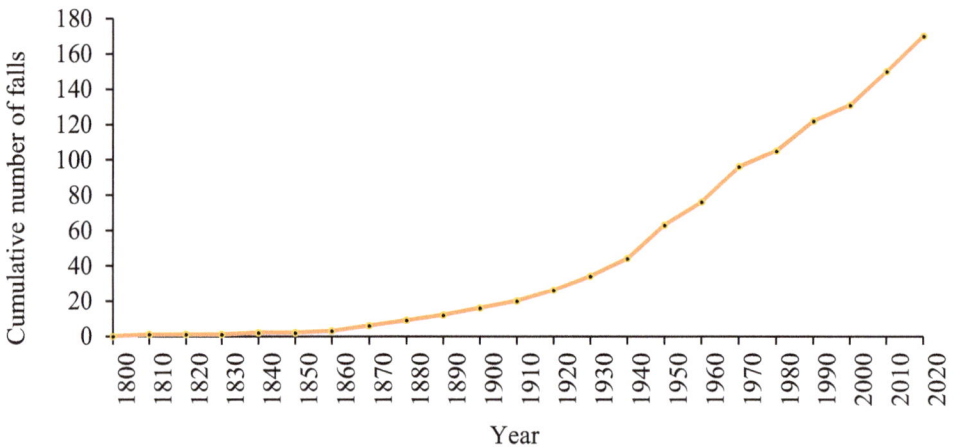

Fig. (1). Cumulative number of observed meteorite falls recorded in Africa.

The evolution of the number of observed meteorite falls every 10 years since 1800 Fig. (**2**) shows a variable temporal distribution. In particular, the falls rate increased from 0.017 meteorites/106 km^2 per 10 years (3 falls only) between 1800-1860 to 0.17 falls/106 km^2/10 years (41 falls) during the period 1860-1940. In particular, a small but uniform increase in the number of recoveries occurred during the period 1900-1940 (2 new records of meteorite falls per decade).

The recovery rate tripled with 126 falls (0.52 falls/106 km²/10 years) recorded between 1940 and 2020), which represent 74.1% of the collection over the entire period 1800-2020. Furthermore, since 1940 a certain periodicity in the distribution of observed meteorite falls occurred every 10 years, *i.e.,* a decade experiencing 10 falls, that is, on average, one fall per year, followed by a decade with twice more recordings (19 falls on average/10 years, *i.e.,* two falls per year). However, the limited number of falls considered in this study (170) would suggest that this result is only a statistical coincidence due to the "statistics of small numbers", as the evolution of the number of all meteorite falls throughout the world over a period of 10 years does not show any trend (Bland *et al.*, 1998).

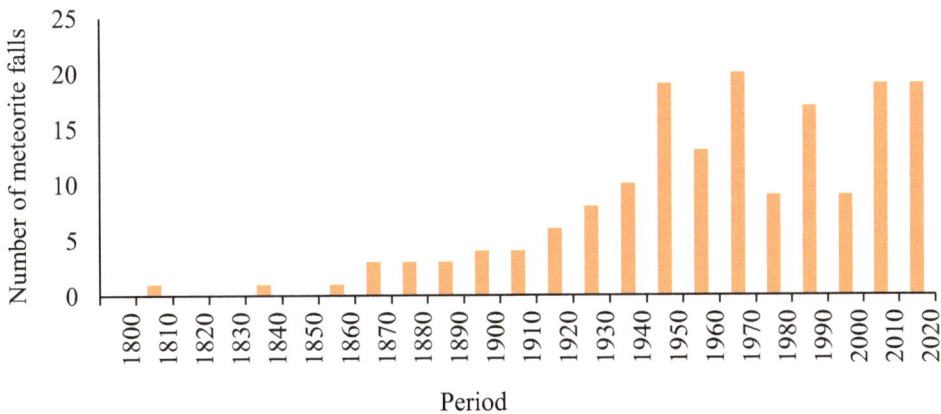

Fig. (2). Evolution of observed meteorite falls number in Africa between 1801 and 2020.

Evolution of Falls Masses

In Africa, the mass of an observed fall ranges from 1.4 g (Natal, South Africa) to 175 kg (Zag, Morocco). The distribution on a logarithmic scale of African meteorite falls according to the retrieved masses Fig. (**3**) shows that the higher number of falls has a mass between 1 and 10 kg (43.75%). A total mass of fragments greater than 100 kg recovered from the same observed meteorite fall is rare (only 3.75%). 21.9% of all African falls consist of fragments totaling a mass between 10 and 100 kg, and almost one-third (30.6%) by a mass of less than 1 kg. Noteworthy, the total mass of fall fragments that are effectively recovered is lower than the actual mass that strikes the ground, as many fragments are being missed due to the difficult observation and recovery of meteorites on the ground.

Furthermore, the chance that a meteorite hits the Earth's ground keeps a mass close to that of when it detaches from the parent body depends on its entry velocity (Heide, 1957). During its atmospheric transit at a speed of 14.2 km/s, a meteor loses a thickness of approximately 7 to 10 cm of its surface layer due to atmospheric ablation (Hughes, 1992). Gounelle (2009) and Koschny *et al.* (2019) reported that tennis ball-sized grains pass through the Earth atmosphere daily, but almost all vaporize completely. Thus, the mass of a collected meteorite fall depends on the size of fragments resulting from its explosion and fallen on the ground. In Slovakia, monitoring the fall of Zvolen on May 27, 1979, allowed to calculate the ratio between the mass of the meteor entering the atmosphere and the mass actually reaching the ground, *i.e.,* a total mass of about 1 kg of recovered fragments versus an incoming mass of 230 kg (Hugues, 1992). However, it is also possible that the inhabitants of the area of the fall did not report the recovered fragments to authorities or scientists. For example, this was the case in Nigeria, where natives did not reveal for years their discovery of Uwet iron fragments (54 Kg) that they found in 1903 (Burke, 1991). This case would explain the low fallen masses recovered in Africa. Thus, the actual masses of African meteorites would be considerably higher than the values shown in Fig. (**3**).

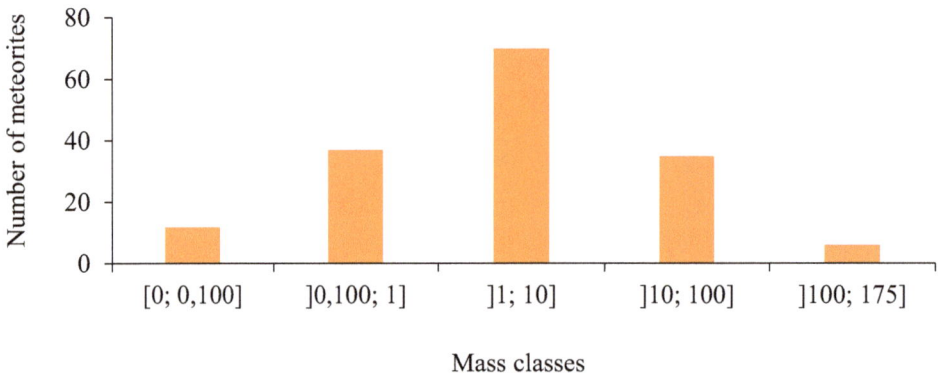

Fig. (3). Distribution of observed meteorite falls in Africa as a function of retrieved mass (kg).

The evolution of masses recovered from meteorites as a function of the century of their fall Fig. (**4**) reveals that the total mass recorded during the 19th century is 81.25 kg, that collected in the entire 20th century is much higher, reaching 1494 kg, and that recovered during the last two decades (2000-2020) reached 844 kg. The latter result would be related to the interest in research and collection of extraterrestrial rock fragments by the African population that has become increasingly large, in addition to the advent of the space interest worldwide, and the consequent increase in research activity in some countries of the African continent.

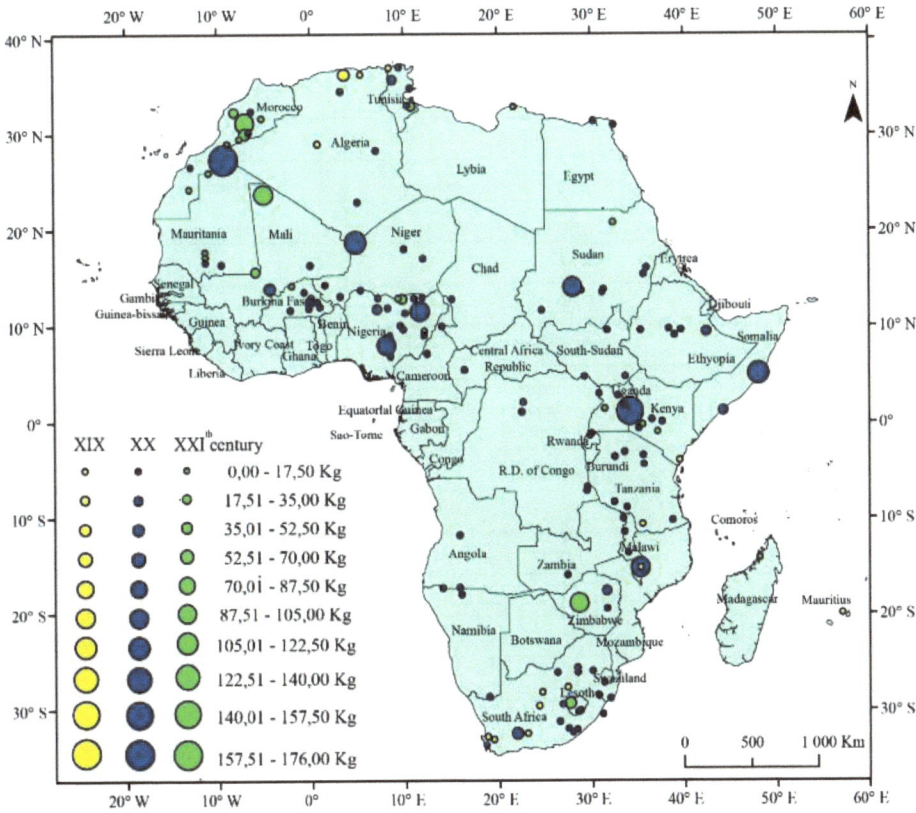

Fig. (4). Distribution of meteorite falls in Africa according to the century of fall.

Evolution of Falls Classes

The processes and results of meteorite classification have evolved since Tschermak's proposal in 1885. Specifically, this study refers to the Weisberg 2006 model that was later developed by Krot *et al.* (2013). Meteorite falls recovered in Africa are mainly composed of silicates (olivines, pyroxenes, feldspars), sulfides and metals, and belong to various classes (Table. **1**), including (i) Chondrites of three types, *i.e.,* carbonaceous, ordinary (H, high iron; L, low iron; LL, low iron and low metal) and enstatites; (ii) Achondrites of only four classes, *i.e.,* Martian, Aubrite, Ureilite and HED (Howardite, Eucrite and Diogenite) types; (iii) iron meteorites; and (iv) unclassified/unknown meteorites. Noteworthy, no Angrite, lunar and stony-iron meteorite falls have been recorded in Africa until December 31, 2020 (refer to Chapter 1 for more details on meteorite classes and subclasses).

Table 1. Types and percentages of different classes of observed meteorite falls in Africa.

Types			Number	%
Chondrites	Carbonaceous		7	4.1
	Ordinary	H	50	29.4
		L	56	32.9
		LL	16	9.4
	Rumuruti		1	0.6
	Enstatite		5	2.9
Achondrites	Martian		3	1.8
	Aubrites		2	1.2
	Urelites		1	0.6
	Angrites		0	0
	HED		11	6.5
	Lunar		0	0
Stony-Iron			0	0
Iron			8	4.7
Unclassified / unknown			10	5.9
-	-	-	**170**	**100**

Data in Table. **1** show that chondrites are the most abundant meteorite falls type recovered in Africa (135, *i.e.,* 122 ordinary chondrites, including [H] (50), [L] (56) and [LL] (16), 7 carbonaceous, 5 enstatites and 1 rumuruti), which account for 79.3% of all observed falls in Africa. The class of achondrites is represented by 17 meteorites, *i.e.,* 10.1% of the total African falls (11 HED, 3 martians, 2 aubrites, 1 urelite). The eight iron meteorites confirm the rarity of this type of observed falls, *i.e.,* 4.7% of the total African falls. Finally, ten, *i.e.,* 5.9%, meteorites are listed as unclassified or uncertain.

The temporal evolution of the types/categories of meteorite falls in Africa (Table. **1**) presented in Fig. (**5**) shows that all observed falls were chondrites during the period 1800-1880, and remained dominant over other types until 2020. Achondrites and iron meteorites have been recorded in Africa only since 1880, but no iron meteorites were recorded after 1988.

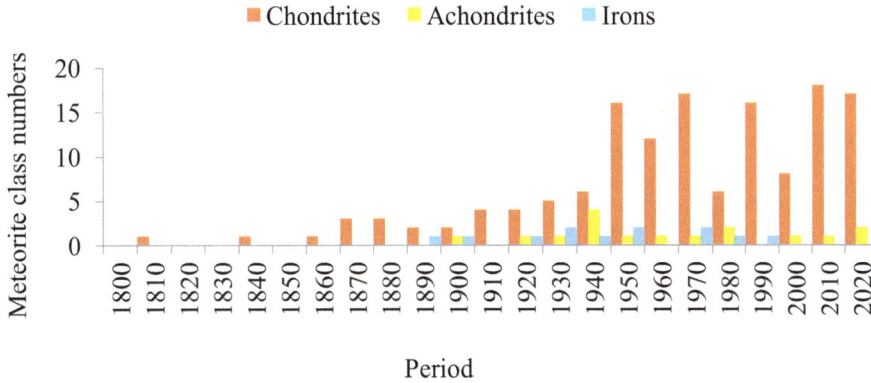

Fig. (5). Observed meteorite falls (chondrites, achondrites and irons) in Africa between 1800 and 2020.

Geographical Distribution of Observed Meteorite Falls in Africa

The number of observed falls that have been recorded in Africa reached 170 during the period 1800-2020, or 0.056 falls/10,000 km². Fig. (4) shows the distribution of African meteorite fall coordinates using the ArcGIS 10.2 application. Currently, more than 83 African meteorite falls (48.8% of total falls) have been recorded in Eastern and Southern Africa, and 43.6% of falls occurred in North West Africa. Most falls occurred in four countries, *i.e.,* South Africa (22), Morocco (19), Nigeria (17) and Sudan (11) (Table. **2**). Northeast (north of latitude 18°N) and Central Africa recorded only a few recoveries, *i.e.,* 7.6% of the African collection.

Table 2. Observed meteorite falls in Africa (data from The Meteoritical Bulletin database, 2020).

Meteorite Fall Name	Country	Year of Fall	Month of Fall	Day of Fall	Mass (kg)	Type	Latitude	Longitude
Aumale	Algeria	1865	August	25	50	L6	36° 10'N	3° 40'E
Tadjera	Algeria	1867	June	9	9	L5	36° 11'N	5° 25'E
Feid chair	Algeria	1875	August	16	0,38	H4	36° 53'N	8° 27'E
Hassi jekna	Algeria	1890	–	–	1,25	Iron-IAB	28° 57'N	0° 49'E
Fort fllatters	Algeria	1944	June	23	–	st-uncl	28° 15'N	7° 0'E
Seldbourak	Algeria	1947	February	26	0,15	H5	22° 50'N	4° 59'E
El idrissia	Algeria	1989	March	10	10	L6	34° 25'N	3° 15'E
Oued Sfayat	Algeria	2019	May	16	8	H5	26°42.4'N	6°9.7'W
Ehole	Angola	1961	August	31	2,4	H5	17° 18'S	15° 50'E
Chitado	Angola	1966	October	20	–	L6	17° 21'S	13° 58'E

(Table 2) cont.....

Meteorite Fall Name	Country	Year of Fall	Month of Fall	Day of Fall	Mass (kg)	Type	Latitude	Longitude
Jolomba	Angola	1974	February	3	0,483	LL6	11° 51'S	15° 50'E
Béréba	Burkina Faso	1924	June	27	18	Eu-mm	11° 39'N	3° 39'W
Nadiabondi	Burkina Faso	1956	July	27	8,17	H5	12°N	1°E
Gao-Guenie	Burkina Faso	1960	March	5	-	H5	11° 39'N	2° 11'W
Bogou	Burkina Faso	1962	August	14	8,8	Iron, IAB	12° 30' N	0° 42 E
Guibga	Burkina Faso	1972	February	26	0,288	L5	13° 30'N	0° 41'W
Pê	Burkina Faso	1989	June	14	-	L6	11° 20' 1''N	3° 32' 32''W
Bilanga	Burkina Faso	1999	October	27	25	Diogenite	12° 27'N	0° 5'W
Gasseltepa oua	Burkina Faso	2000	August	22	-	H5	14° 9' 3''N	2° 2' 30''W
Ouadagou	Burkina Faso	2003	November	-	4,44	L5	12° 54' 0''N	0° 4' 48''E
Guideer	Cameroun	1949	January	7	0,968	LL5	9° 55'N	13° 59'E
Galim (a)	Cameroun	1952	November	13	0,0361	LL6	7° 3'N	12° 26'E
Galim (b)	Cameroun	1952	-	-	0,028	EH3/4-an	7° 3'N	12° 26'E
Bali	Central African Republic	1907	November	22	1	CV3	5° 23'N	16° 23'E
Djermaia	Chad	1961	February	25	3,95	H	12° 44'N	15° 3'E
Alberta	Congo-Dem-Rep	1949	November	13	0,625	L	2° 0'N	22° 40'E
Lusaka	Congo-Dem-Rep	1951	October	31	0,95	H4	6° 50'S	29° 30'E
Yambo	Congo-Dem-Rep	1951	April	-	-	Unknown	7° 13'S	29° 26'E
Kalaba	Congo-Dem-Rep	1951	October	20	0,004	H5	1° 0'N	22° 30'E
Essebi	Congo-Dem-Rep	1957	July	28	0,5	C2-ung	2° 53'N	30° 50'E
Nakhla	Egypt	1911	June	28	10	Martian	31° 19'N	30° 21'E
Sinai	Egypt	1916	July	14	1,455	L6	30° 54'N	32° 29'E
Holetta	Ethiopia	1923	April	14	1,415	St-uncl	9° 4'N	38° 25'E
Ankober	Ethiopia	1942	July	7	6,5	H4	9° 32'N	39° 43'E
Wolamo	Ethiopia	1964	August	-	0,1664	OC	9°N	39°E
Nejo	Ethiopia	1970	May	11	2,45	L6	9° 30'N	35° 20'E
Gursum	Ethiopia	1981	February	10	34,65	H4/5	9° 22'N	42° 25'E
Bawku	Ghana	1989	December	29	1,557	LL5	11° 5'N	0° 11'W
Daoukro	Ivoary Coast	-	-	-	-	Doubtful	7° 5'N	3° 58'W
Duroma	Kenya	1853	March	6	0,577	L6	4° 0'S	39° 30'E
Rumuruti	Kenya	1934	January	28	0,067	R3,8-6	0° 16'N	36° 32'E
Meru	Kenya	1945	Februry	2	6	LL6	0° 0'N	37° 40'E
Sotik	Kenya	1946	-	-	2,05	Doubtful	0° 41' S	35° 7' E
Thika	Kenya	2011	September	26	14,2	L6	1° 0' 10''S	37° 9' 1''E
Kuresoi	Kenya	2014	February	27	0,555	L6	0° 18' 0''S	35° 31' 44''E
Thuate	Lesotho	2002	July	21	45,3	H4/5	29° 20'S	27° 35'E

(Table 2) cont.....

Meteorite Fall Name	Country	Year of Fall	Month of Fall	Day of Fall	Mass (kg)	Type	Latitude	Longitude
Werdama	Libya	2006	May	21	4	H5	32° 47' 50"N	21° 47' 14"E
Maromandi	Madagascar	2002	July	5	6	L6	14° 12'S	48° 6'E
Benenitra	Madagascar	2018	July	27	20	L6	-23°26.7'S	-45°4.7'E
Zomba	Malawi	1899	January	25	7,5	L6	15° 11'S	35° 17'E
Mtola	Malawi	1944	June	17	1,1	stone-uncl	11° 30'S	33° 30'E
Dowa	Malawi	1976	March	25	0,642	stone-uncl	13° 40'S	33° 55'E
Machinga	Malawi	1981	January	22	93,2	L6	15° 12' 44"S	35° 14' 32"E
Chisenga	Malawi	1988	January	17	3,92	Iron, IIIAB	10° 3' 34"S	33° 23' 42"E
N'Goureyma	Mali	1900	June	5	37,5	Iron, ung	13° 51'N	4° 23'W
Gao (Mali)	Mali	1932	–	–	–	Doubtful	16° 18'N	0° 0'E
Chergach	Mali	2007	July	2	100	H5	23° 41' 47"N	5° 0' 53"W
Kiffa	Mauritania	1970	October	23	1,5	H5	16° 35'N	11° 20'W
Aioun el Atrouss	Mauritania	1974	April	17	1	Diogenite-pm	16° 23' 53"N	9° 34' 13"W
Boumdeid 2003	Mauritania	2003	September	24	0,19	L6	17° 42' 38"N	11° 22' 17"W
Basikounou	Mauritania	2006	October	16	29,56	H5	15° 47'N	5° 54'W
Boumdeid 2011	Mauritania	2011	September	14	3,6	L6	17° 10' 30"N	11° 20' 29"W
Mreira	Mauritania	2012	December	–	6	L6	25° 57' 33"N	10° 57' 37"W
Mhabes el Hamra	Mauritania	2018	December	23	23.1	H4/5	26.2269568°N	7.8181461°W
Mauritius	Mauritius	1801	Before December	–	0,22	L6	20°S	57°E
Douar Mghila	Morocco	1932	August	20	1,161	LL6	32° 20'N	6° 18'W
Ouled el Hadjar	Morocco	1986	April	–	1,216	LL6	30° 10' 48"N	6° 34' 38"W
Itqiy	Morocco	1990	–	–	4,72	EH7-an	26° 35' 27"N	12° 57' 8"W
Zag	Morocco	1998	August	4	175	H3/6	27° 20'N	9° 20'W
Bensour	Morocco	2002	February	11	45	LL6	30°N	7°W
Oum Dreyga	Morocco	2003	October	16	17	H3/5	24° 18'N	13° 6'W
Benguerir	Morocco	2004	November	22	25	LL6	32° 15' N	08° 09' W
Tamgakht	Morocco	2008	December	20	100	H5	31° 9' 48"N	7° 0' 54"W
Tissint	Morroco	2011	July	18	7	Martian	29° 28' 55"N	7° 36' 40"W
Tirhert	Morroco	2014	July	9	8	Eucrite	28° 56' 6"N	8° 54' 18"W
Tinejdad	Morroco	2014	September	9	1,86	H5	31° 36' 33"N	5° 11' 39"W
Sidi Ali Ou Azza	Morocco	2015	July	28	1,500	L4	29°47'2.9"N	7°23'21.8"W
Oudiyat Sbaa	Morocco	2016	November	18	23,85	EH5	25.546°N	12.418°W
Matarka	Morocco	2018	January	5	0,538	L6	33°15'N	02°45'W
Gueltat Zemmour	Morocco	2018	August	21	8	L4	25°05'32.0"N	12°37'23.9"W
Ksar El Goraane	Morocco	2018	October	28	4	H5	32°11'13.2"N	003°42'04.6"W

(Table 2) cont.....

Meteorite Fall Name	Country	Year of Fall	Month of Fall	Day of Fall	Mass (kg)	Type	Latitude	Longitude
Wad Lahteyba	Morocco	2019	June	27	20	H5	27°22'23.153"N	8°58'51.744"W
Al Farciya	Morocco	2019	August	20	01h15'	1,3	L6	27°01'27.59"N
Tarda	Morocco	2020	August	25	4	C2-ung	31° 49' 35"N	4° 40' 46"W
Ovambo	Namibia	1900	–	–	0,1215	L6	18°S	16°E
Witsand Farm	Namibia	1932	December	1	0,066	LL4	28° 40'S	18° 55'E
Birni N'Konni	Niger	1923	April	–	0,56	H4	13° 46'N	5° 18'E
Aïr	Niger	1925	–	–	24	L6	19° 5'N	8° 23'E
Dosso	Niger	1962	February	19	1,25	L6	13° 3'N	3° 10'E
Niger (L6)	Niger	1967	August	1	0,0033	L6	18	10
Niger (LL6)	Niger	1967	August	1	0,0033	LL6	17	12
Tillaberi	Niger	1970	–	–	3	L6	14° 15'N	1° 32'E
Mount Tazerzait	Niger	1991	August	21	110	L5	18° 42'N	4° 48'E
Pétèlkolé	Niger	1995	April	10	0,189	H5	14° 3' 7"N	0° 25' 12"E
Maigatari-Danduma	Niger	2004	August	1	4,63	H5/6	12° 50'N	9° 23'E
Udei Station	Nigeria	1927	Spring	–	103	Iron, IAB	7° 57'N	8° 5'E
Bununu	Nigeria	1942	April	–	0,357	Howardite	10° 1'N	9° 35'E
Git-Git	Nigeria	1947	January	9	0,48	L6	9° 36'N	9° 55'E
Karewar	Nigeria	1949	September	19	0,18	L6	12° 54'N	7° 9'E
Geidam	Nigeria	1950	July	6	7,25	H5	12° 55'N	11° 55'E
Akwanga	Nigeria	1959	July	2	3	H	8° 55'N	8° 26'E
Zagami	Nigeria	1962	October	3	18	Martian	11° 44'N	7° 5'E
Ohuma	Nigeria	1963	April	11	7,7	L5	6° 45'N	8° 30'E
Kabo	Nigeria	1971	April	25	13,4	H4	11° 51'N	8° 13'E
Mayo Belwa	Nigeria	1974	August	3	4,85	Aubrite	8° 58'N	12° 5'E
Gashua	Nigeria	1984	April	–	4,16	L6	12° 51'N	11° 2'E
Gujba	Nigeria	1984	April	3	100	Cba	11° 29' 30"N	11° 39' 30"E
Katagum	Nigeria	1999	September	–	1,5	L6	11° 20'N	10° 5'E
Kilabo	Nigeria	2002	July	21	19	LL6	12° 46'N	9° 48'E
Demsa	Nigeria	2006	October	13	5	H6	9° 27' 21"N	12° 9' 9"E
Aba Panu	Nigeria	2018	April	19	160	L3	8°16'55.83"N	3°34'1.72"E
Ruhobobo	Rwanda	1976	October	21	0,466	L6	1° 27'S	29° 50'E
Bur-Gueluai	Somalia	1919	October	16	120	H5	5°N	48°E
Ergheo	Somalia	1989	July	-	20	L5	1° 10'N	44° 10'E
Cold Bokkeveld	South Africa	1838	October	13	5,2	CM2	33° 8'S	19° 23'E
Daniel's Kuil	South Africa	1868	March	20	1,064	EL6	28° 12'S	24° 34'E
Cronstad	South Africa	1877	November	19	3,65	H5	27° 42'S	27° 18'E
Wittekrantz	South Africa	1880	December	9	2,2	L5	32° 30'S	23° 0'E

(Table 2) cont.....

Meteorite Fall Name	Country	Year of Fall	Month of Fall	Day of Fall	Mass (kg)	Type	Latitude	Longitude
Piquetberg	South Africa	1881	–	–	0,037	H	32° 52'S	18° 43'E
Orange River (stone)	South Africa	1887	–	–	0,008	Doubtful	29° 40'S	24° 13'E
Jackasfont	South Africa	1903	April	22	48	L6	32° 30'S	21° 54'E
St, Mark's	South Africa	1903	January	3	13,78	EH5	32° 1'S	27° 25'E
Diep River	South Africa	1906	November	4	1	L6	33° 45'S	18° 34'E
N'Kandhla	South Africa	1912	June	21	0,46	L6	25° 40'S	28° 22'E
Leeuwfonte	South Africa	1912	August	1	17,2	Iron, IID	28° 34'S	30° 42'E
Witklip Farm	South Africa	1918	May	26	0,022	H5	26°S	30°E
Maria Linden	South Africa	1925	April	15	0,114	L4	30° 15' S	28° 28' E
Queen's Mercy	South Africa	1925	April	30	7	H6	30° 7'S	28° 42'E
Malvern	South Africa	1933	November	20	0,807	Eu-pm	29° 27'S	26° 46'E
Macibini	South Africa	1936	September	23	1,995	Eu-pm	28° 50'S	31° 57'E
Benoni	South Africa	1943	July	25	3,88	H6	26° 10'S	28° 25'E
Molteno	South Africa	1953	May	–	0,15	Howardite	31° 15'S	26° 28'E
Idutywa	South Africa	1956	February	1	3,46	H5	32° 6'S	28° 20'E
Bulls Run	South Africa	1964	–	–	2,25	Iron	32° 24 S	27° 52'E
Lichtenberg	South Africa	1973	September	26	4	H6	26° 9'S	26° 11'E
Natal	South Africa	1973	–	–	0,0014	St-uncl	29° 81'S	30° 83'E
Maridi	South Sudan	1941	–	–	3,2	H6	4° 40'N	29° 15'E
Kapoeta	South Sudan	1942	April	22	11,36	Howardite	4° 42'N	33° 38'E
Malakal	South Sudan	1970	August	10	2	L5	9° 30'N	31° 45'E
Khor Temiki	Sudan	1932	April	8	3,2	Aubrite	16°N	36°E
Umm Ruaba	Sudan	1966	December	27	1,7	L5	13° 28'N	31° 13'E
Kingai	Sudan	1967	November	7	0,0674	H6	11° 38'N	24° 41'E
Kidairat	Sudan	1983	January	–	100	H6	14°N	28°E
Umm Ruaba(b)	Sudan	1983	–	–	—	L4	13° 46'N	31° 21'E
New Halfa	Sudan	1994	November	8	12	L4	15° 22'N	35° 41'E
Alzarnkh	Sudan	2001	February	8	0,7	LL5	13° 39' 37"N	28° 57' 36"E
Almahata Sitta	Sudan	2008	October	7	3,95	Urellite-an	20° 44' 45"N	32° 24' 46"E
Dwaleni	Swaziland	1970	October	12	3,23	H4/6	27° 12'S	31° 19'E
Peramiho	Tanzania	1899	October	24	0,165	Eu-mm	10° 40'S	35° 30'E
Malampaka	Tanzania	1930	September	–	0,47	H	3° 8'S	33° 31'E
Ivuna	Tanzania	1938	December	16	0,705	CI1	8° 25'S	32° 26'E
Rupota	Tanzania	1949	February	7	6	L4/6	10° 16'S	38° 46'E
Ishinga	Tanzania	1954	October	8	1,3	H	8° 56'S	33° 48'E
Ufana	Tanzania	1957	August	5	0,1892	EL6	4° 26'S	35°35'E

(Table 2) cont.....

Meteorite Fall Name	Country	Year of Fall	Month of Fall	Day of Fall	Mass (kg)	Type	Latitude	Longitude
Karatu	Tanzania	1963	September	11	2,22	LL6	3° 30'S	35° 35'E
Chela	Tanzania	1988	July	12	2,94	H4	3° 40'S	32° 30'E
Tatahouine	Tunisia	1931	June	27	12	Diogenite	32° 57'N	10° 25'E
Dahmani	Tunisia	1981	May	-	18	LL6	35° 37'N	8° 50'E
Sfax	Tunisia	1989	October	16	7	L6	34° 45'N	10° 43'E
Djoumine	Tunisia	1999	October	31	10	H5/6	36° 57'N	9° 33'E
Beni M'hira	Tunisia	2001	-	8	19	L6	32° 52'N	10° 48'E
Maziba	Uganda	1942	September	24	4,98	L6	1° 13'S	30° 0'E
Soroti	Uganda	1945	-	17	2,05	Iron, ungrouped	1° 42'N	33° 38'E
Awere	Uganda	1968	July	12	0,134	L4	2° 43'N	32° 50'E
Mbale	Uganda	1992	August	14	150	L5/6	1° 4'N	34° 10'E
Hoima	Uganda	2003	March	30	0,1677	H6	1° 20' 42"N	31° 28' 22"E
Monze	Zambia	1950	October	5	-	L6	15° 58'S	27° 21'E
Mangwendi	Zimbabwe	1934	March	7	22,3	LL6	17° 39'S	31° 36'E
Magombed	Zimbabwe	1990	July	2	0,667	H3/5	19° 29'S	31° 39'E
Nkayi	Zimbabwe	2009	March	1	100	L6	18° 56'S	28° 36'E

MAJOR FACTORS AFFECTING THE OBSERVATION AND COLLECTION OF METEORITE FALLS IN AFRICA

In this study, besides the factors used in the literature, *i.e.,* population density and historical factors, other geographical and human factors that could contribute to the recovery of meteorite falls in Africa are considered, given the diversity of its environment. Demographic data were obtained from various FAO reports.

Observed Meteorite Fall Rate and Population Density

In this section, the variation in the number of observed meteorite falls in Africa is related to the variation in population density in the continent. In 1700, Africa counted 100 million inhabitants, *i.e.,* 17% of the world's population. After 200 years, *i.e.,* in 1900, this continent still maintained the same population, *i.e.,* only 6% of the world population. During these two centuries, the effects of the slave trade (35 million slaves deported from Africa), European colonial penetration and internal wars depopulated the continent, causing a marked demographic recession (Louise and Diop, 1985). The relative decrease in the African population would be one factor among others that would explain the low rate of observed meteorite falls and recoveries during the 19th century in the continent.

Population growth resumed during the period 1920-1985, passing from 2.2% to 2.8%. In particular, the population density in Africa increased from 8 inhabitants per km² in 1950 to 45 inhabitants per km² in 2020. Currently, the population density in Africa is ranked second after that of Asia.

Fig. (**6**) shows the significant correlation (coefficient of determination r = 0.98) that exists between the variation of the cumulative number of observed meteorite falls in Africa and the population density in this continent from 1950 to 2020. During this period (70 years), the population density in Africa increased by 560%, and 107 African meteorite falls were recorded, *i.e.,* observed falls increased by 270%. Between 1950 and 1990, the meteorites fall rate varied roughly linearly with population density, whereas during the period 1990-2020, a deficit of observed and recovered falls occurred, as about 115 meteorite falls could have been expected reasonably during the last 3 decades. Thus, the rate of observed meteorite falls in Africa is clearly related, among other factors, to the growth of population density.

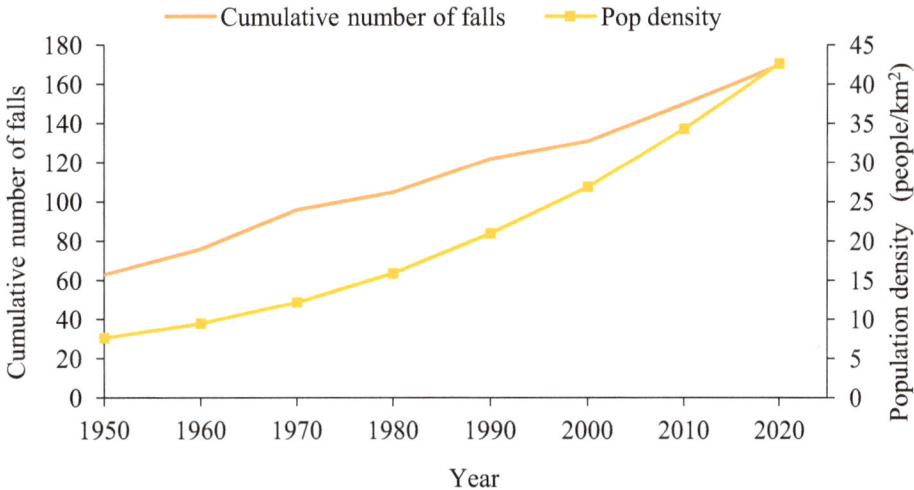

Fig. (6). The cumulative number of meteorite falls recorded and the population density (people per square kilometer) in Africa from 1950 to 2020.

The geographical distribution of observed meteorite falls in Africa related to the population density in each African country is illustrated in Fig. (**7**). Apparently, most meteorites have been recorded in regions with high or medium population density (northern Morocco, Algeria and Tunisia, Nigeria, Burkina-Faso, Uganda, western Kenya and south-eastern South Africa). Despite its low population density, which does not exceed five people per km², a few observed meteorite falls have been recorded in the western part of the Sahara Desert. However, this region would result much more populated if the population density is related to the km of roads and trails, as up to 600 inhabitants/km are present along the main roads in some Algerian towns with more than 10,000 inhabitants, such as Tamanrasset, Arak and Reggane (SWAC, 2015). Today most people in the Sahara live in villages or move as nomads from region to region across the desert.

Considering the different countries of the African continent, the rate of meteorite falls increases with their population density. However, the population distribution is unequal in Africa, *i.e.,* some countries featuring a population condensed in one part of the country, such as Libya and Egypt (in the North or on the banks of the Nile, respectively), recorded only a few falls (2 falls only). Thus, the uneven distribution of population in a country would make it difficult to observe and recover meteorite falls. For example, in Western Australia, meteorites are rarely seen as they fall due to a sparse and unevenly distributed population. Only 13 observed falls were recorded during the period 1854-1992 in this region of Australia (Bevan, 1992), a number that makes the falls/findings ratio low, *i.e.,* not exceeding 1:20. This result confirms the importance of the uniform distribution of inhabitants in a country for increasing the rate of observed meteorite falls. Other studies have also confirmed that the distribution of human populations affects the probability of perception and retrieval of meteorites (Wickman and Palmer, 1979 and Rasmussen, 1991).

In this study, the variation of the number of observed meteorites falls with that of the total population, the rural population, and the urban population (averages for the period 1990-2014) has been compared among different African countries. Results show that the number of observed meteorite falls correlated more significantly with the rural population ($r = 0.69$), compared to the total population ($r = 0.62$) and the urban population ($r = 0.44$). The stronger correlation with the rural population can be explained by its lifestyle, in particular, that of nomads, who constantly live in contact with nature and have developed a meteorite know-how that enables them to observe and search for them. These people can observe meteors falling and describe to scientists about this phenomenon, while in urban areas, people generally miss the occurrence of many possible falls (Ibhi, 2016).

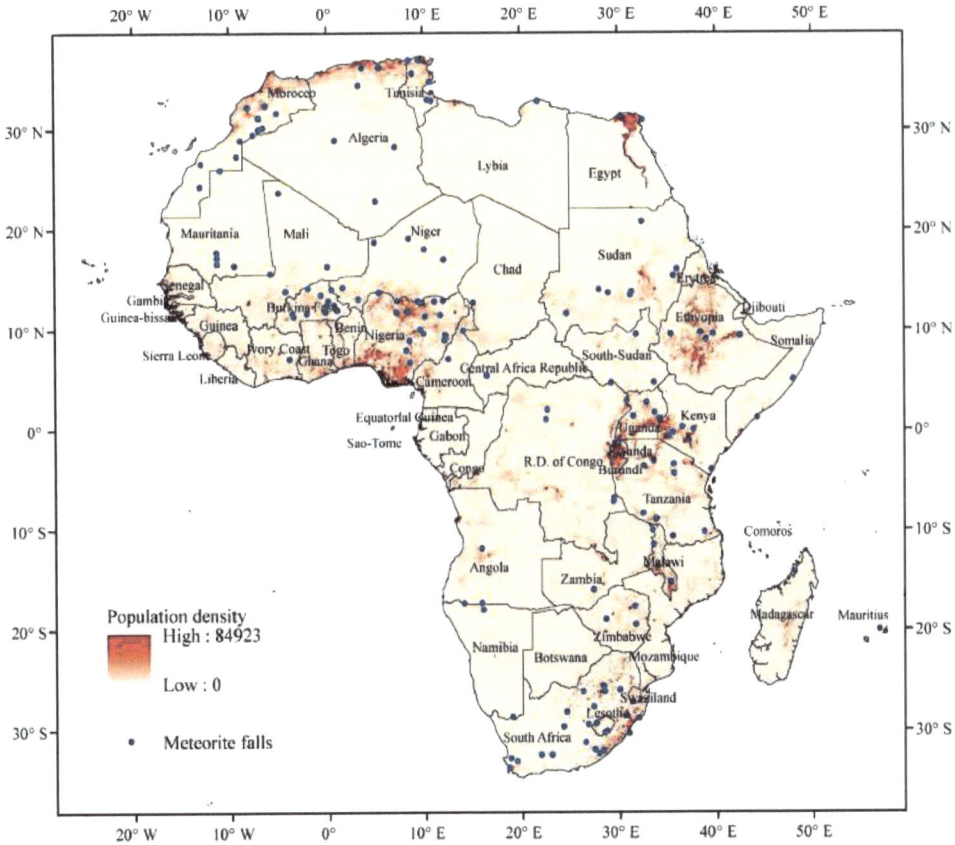

Fig. (7). Geographical distribution of meteorite falls in Africa as a function of population density.

In conclusion, the fact that the rate of observed falls has not followed the increase in population density since 1990 (Figure **5**) could be related to the migration of the rural population to cities. The evolution of the average rural population in some African countries supports this explanation, *i.e.,* between 1990 and 2018, it decreased from 66.6% to 28.4% in Algeria, from 48% to 33% in South Africa, from 48% to 38% in Morocco, and from 73% to 49.7% in Nigeria.

Observed Meteorite Falls Rate and Historical and Educational Factors

The distribution of observed meteorite falls in Africa according to the century of their fall is illustrated in Fig. (**4**), which shows that the rate was low in the 19th century (14 falls only), and was eight times higher during the 20th century, which features 117 meteorite falls. Then, 39 meteorites were observed and listed between 2000 and 2020.

The low rate of falls recorded between 1800 and 1860 may also be related to the lack of knowledge and education in the field of meteorites. In ancient times, the passage of meteors through the atmosphere was associated with superstitions such as the death of a famous individual or the passage of the devil among the inhabitants of many African countries. Even in the United States of America until 1984, these falls were valued upon arrival, but not considered valuable items to acquire (Clarke *et al.*, 2006).

The slight increase recorded between 1860 and 1920 can be related to the arrival of Europeans dating from 1860 in Algeria and South Africa, which are among the first African countries to have listed observed meteorite falls in the 19th century (Fig. **4**). Noteworthy, the most successful African countries in terms of perception and recovery of meteorite falls were colonized by Great Britain (South Africa, Nigeria, Sudan and Malawi) and France (Morocco, Algeria and Tunisia). Moreover, in most African countries, the first observed meteorite falls are dated after their colonization, for example, Algeria documented its first fall in 1865, and Malawi in 1899. Both Great Britain and France have experienced meteorite falls since the early 17th century, and many falls were observed and collected during the 19th century in these two countries. According to Dodd (1986), France was the first country to recognize the scientific importance of meteorites around 1800. The colonization of African countries increased in 1885 and became widespread in 1914 in parallel with population growth, which would have affected the increasing rate of observed meteorite falls in the 20th century. From this analysis emerges that colonization had a cultural impact on meteorite education in some African countries.

Dodd (1986) and Bevan (1992) asserted that historical and educational factors have also contributed to the recovery of meteorites in Australia. In Africa, the period 1940-2020 has been very productive in observed falls due to the interest in space exploration and the expansion of research teams interested in studying these alien rocks. In some African countries, meteorites are provided by nomads and local populations who have travelled the deserts in search of fragments to sell. According to Beech (2003), recovering more meteorites requires both an educated and inspired public and an active team of researchers in the field. In this context, activities to raise awareness of the scientific and economic value of meteorites have occurred in certain regions of Africa. In particular, in Morocco, several expeditions were conducted by the two teams of researchers specialized in the field of meteorites from the University Ibn Zohr, Agadir, and the University Hassan II, Casablanca, in areas of meteorite falls such as Tighert, Tissint and Tamdakht. Furthermore, the creation in 2016 of the "Meteorites Museum" at the Ibn Zohr University of Agadir has largely contributed to popularizing meteorite science and training of researchers and collectors of these precious rocks. The

Senegalese Association for the Promotion of Astronomy also aims to popularize science in this field by organizing observations, workshops, conferences, and training. The South African meteorite recovery program is another example that aims to improve the way amateurs can help researchers to acquire and study meteorites (Gould, 2003).

Observed Meteorite Fall Rates as a Function of Surface Size, Forest Cover and Vegetation Type

As the size of the country increases its chances of hosting meteorite falls, many large African countries such as South Africa and Nigeria have been able to list a good number of observed meteorite falls, while most of the small African countries (Sierra Leone, Liberia, Guinea, Benin, Togo, and Burundi) have not reported any fall to date, despite their high population densities. The recorded fall rate also remains very low in Libya, Egypt and Chad and is null in other African countries, despite the large desert areas they have.

With regards to the area of an African country, the rate of observed meteorite falls in countries with large areas such as South Africa, Nigeria, and the Democratic Republic of Congo (respectively 0.18 falls/10,000 km^2, 0 .17 falls/10,000 km^2, 0.02 falls/10,000 km^2) are lower than in some other countries with a smaller surface area, such as Morocco (0.27 falls/10,000 km^2), Tunisia (0.3 falls/10,000 km^2) and Burundi (0.33 fall/10,000 km^2). However, if only the area without forest cover is considered, the rate of observed falls doubles in the Democratic Republic of Congo, for example. Actually, in countries covered by dense tropical forests over large areas, such as Gabon, Congo and the Central African Democratic Republic (65% to 85% of the total area of the country), very few or no falls have been documented. A large forest cover imposes, especially in small countries, the concentration of the population outside the forest areas and consequently, several falls would be missed. Forest cover is, therefore, an unfavorable factor affecting negatively the perception of meteorite falls and their collection by inducing an uneven distribution of inhabitants across several African countries, such as Cameroon and Benin, even with a high population density.

Figs. (**8** and **9**) relate the places of observed meteorite falls in Africa to their vegetation type. Results of this analysis show that 60% of falls were recorded in grassy savannah, 9.5% in desert brush and 6.3% in zones with little or no vegetation, whereas the lowest numbers of meteorites were recorded in forest areas. These results suggest that the type of vegetation is another factor that influences the observation of meteorite falls and their collection. Sparse and low vegetation does not represent an obstacle to the perception of meteors and the collection of meteorites. Noteworthy, despite the presence of savannah in other areas of Africa, as is the case for large parts of central Africa and southwestern West Africa, for example, few or no falls were recorded.

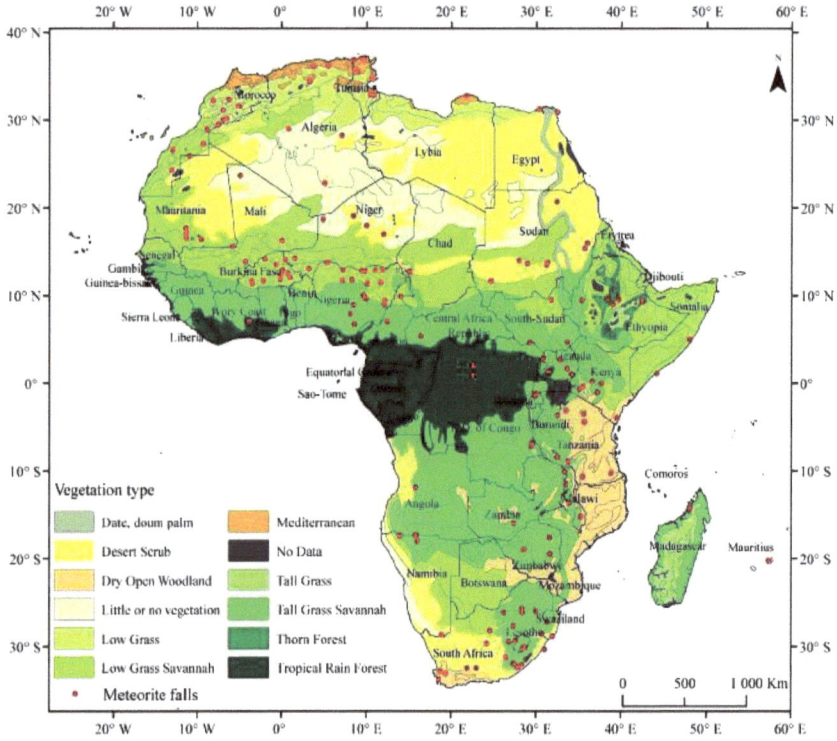

Fig. (8). Geographical distribution of meteorite falls in Africa according to the type of vegetation.

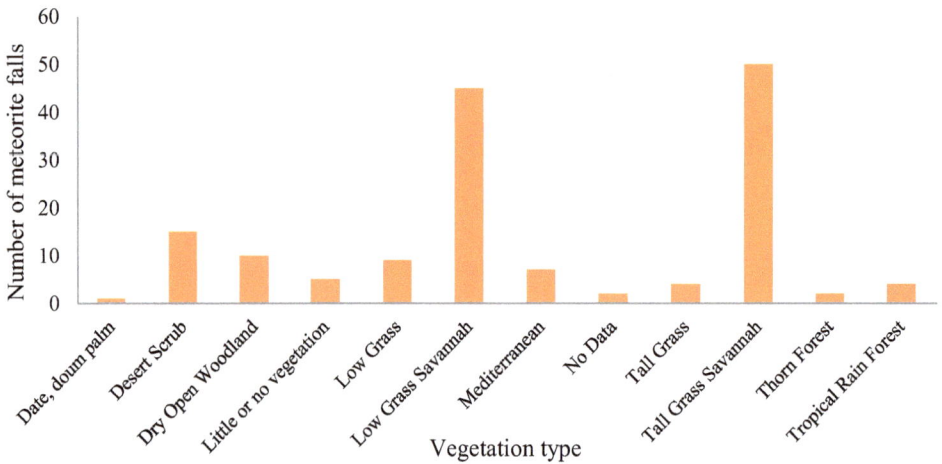

Fig. (9). Number of observed meteorite falls in Africa as a function of the vegetation type.

Observed Meteorite Falls Rate and Distance of their Location from the Road

Fig. (**10**) illustrates the distribution of the road network and the locations of meteorite falls in Africa, showing that the road network does not cover uniformly the entire African territory, *i.e.,* its density is lower in the desert and forest regions than in the proximity of towns and villages. The distribution of the places of observed meteorite falls shows that 45.5% of them were recorded near a road (*i.e.,* at a distance of less than 10 km from the road) and where the population is concentrated, so the perception and discovery of a fall are easier. 40.6% of observed African falls were recorded at a distance of 10 to 50 km from the roads, and only 13.9% of falls were reported in places more than 50 km far from roads. These results show that the low rate of falls recorded in the African continent (0.026 per year for 106 km²) is related to the difficulty of access for meteorite researchers to the fall places, which can delay their discovery or lead to their loss in the wild. On the other hand, results confirm that the rural population, especially in countries that have registered a good number of observed falls, such as South Africa, Nigeria and Morocco, is very interested in extraterrestrial rocks and does not hesitate to travel several km to reach the place of the fall.

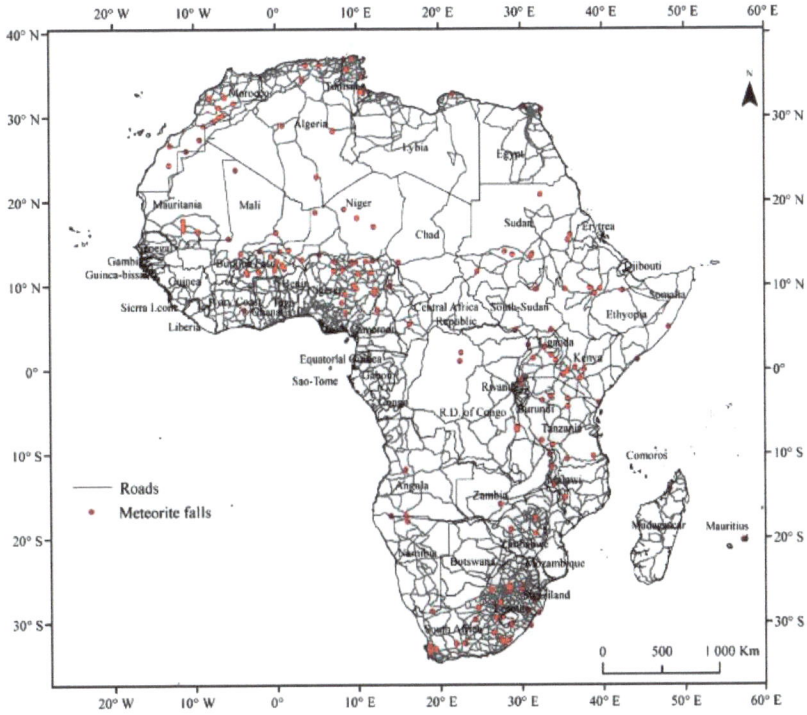

Fig. (10). Geographical distribution of meteorite falls in Africa in relation to the distance from the roads.

In some cases, the meteorite fell near the inhabitants, as for the 18-kg Zagami fall that crashed in Nigeria on October 3, 1962, a few dozen meters from a farmer while he was working in his field. Other meteorites fell far from people who observed their fall, and their recovery needed several days or even months. In Errachidia, in the south-east of Morocco, the observed Eucrite Tighert fall that dates July 9, 2014, was discovered the day after its fall, because it was close to populated areas, whereas the first fragments of the observed Martian fall Tissint, which fell on July 18, 2011, in the Tata-Morocco region, were found only 6 months later with hundreds of nomads of these Saharan regions participating in their search (Ibhi, 2013). Similarly, several weeks of searching were needed to locate the fall of Tamdakht that fell on December 20, 2008 (Chennaoui *et al.* 2009). In the case of the Nzala meteorite, which fell on November 13, 2009, many people from Erfoud, Errachidia and Errich (Morocco) observed the meteor and heard three sonic booms, but the search of meteorite hunters conducted for weeks resulted only in recovering pieces of about 100 and 700 g (Ibhi, 2013b).

The Effect of Climate on Meteorites Falls Observation

In Africa, the climate ranges from humid equatorial to seasonally arid and tropical and to subtropical climates of the Mediterranean type that prevail in the extreme north and south of the continent. Besides these regions, we find the subtropical Sahara in the north and the deserts of Namibia and Kalahari in the south. About 66% of the African continent features an arid or semi-arid climate. Similarly, the distribution of precipitation varies in space and time: it is highest in the countries of Central Africa and Madagascar, while the countries of North Africa receive very low amounts of rain.

In equatorial regions, where observed and recorded meteorite falls are rare or none, tropical forest reigns and rainfall exceeds 1000 mm/year on average (1338 mm/year in Central Africa; 1543 mm/year in the Democratic Republic of Congo; 1830 mm/year in Gabon; and 1646 mm/year in Congo Brazzaville). Heavy rains and cloudy weather complicate the observation of meteorites during their fall and the access to the place of the fall, as well as their recovery in the forest. Moreover, almost all desert countries feature a record of relatively considerable recovery rates, thus, the observation of meteors is favored by the clear sky in arid and semi-arid zones where more than 40% of the African population is settled (Ingram *et al.*, 2002).

The monthly variation of the occurrence of African meteorites whose month of fall is well known (142 observed meteorites) Fig. (**11**) shows that the maxima are recorded in April, July, August and October (between 16 and 19 falls), and the minima during the other months.

Furthermore, the rate of falls between the two wet and dry periods in African countries of the two hemispheres based on the month corresponding to each of the 142 observed African falls Fig. (**12**) shows clearly that the number of observed meteorite falls is higher (60% of falls) during the dry period (especially during summer in the North and spring in the South) compared to that recorded during the wet period. The dry seasons are, therefore a period favorable to observation and recovery thanks to the climatic conditions they provide, *i.e.,* clear skies and people dispersed in the wild. Cooke *et al.* (2015) consider that, due to the often-bad winter conditions in the northern hemisphere, the chance to observe meteors is generally not good, even with spectroscopic video cameras. Differently, March 2015 in Italy was favorable for meteor shower watchers, favoring a better total meteor record in the first half of that year (Molau, 2015). Similarly, according to Bevan (2014), fireballs are frequently reported in the southern part of Western Australia, which enjoys clear skies for an average of 243 days per year.

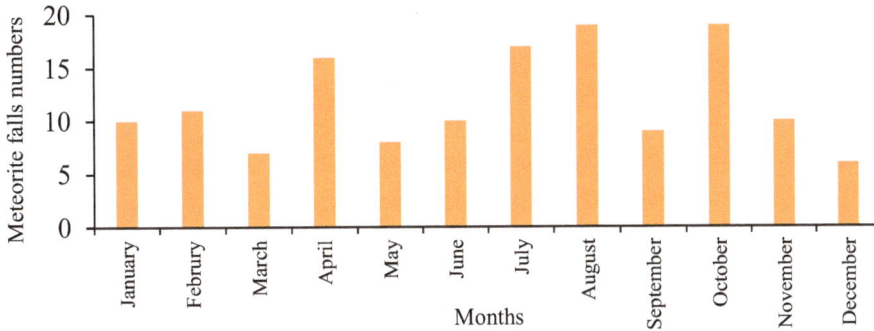

Fig. (11). Variation of observed meteorites numbers in Africa counted per month.

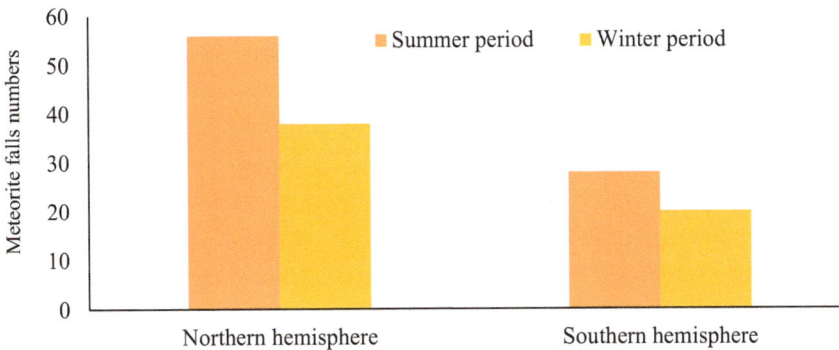

Fig. (12). Comparison of the rate of observed falls between wet and dry periods in African countries of the two hemispheres.

The local time distribution of the 64 observed meteorite falls in Africa Fig. (**13**) reveals that a much higher number of observed falls, *i.e.,* 52 out of 64 falls (81%) occur between 10:00 am and 11:00 pm, with respect to those occurring between 11:00 pm and 10:00 am This result may be related to the presence of people awake during day time and early evening, which increases the chance of observing more meteorite falls, with respect to falls that take place during the night. Similarly, the majority of Australian meteorite falls have been observed by farmers working in the fields during the day (Wetherill, 1968). Assuming that the flow of meteorite falls is the same during day and night, the same number of 52 falls instead of 12 (less than a quarter of the expected ones) should have been observed at night, which implies that more than 75% of meteorite falls are undetected during night.

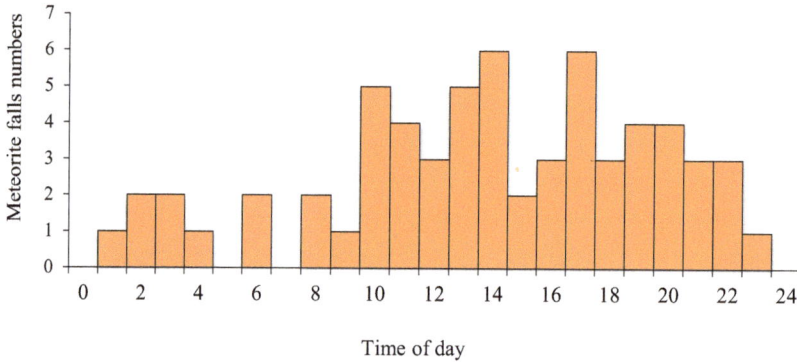

Fig. (13). Distribution of the local time of the fall for 64 African meteorites.

SUMMARY AND DISCUSSION

Without counting the 10 unclassified or unknown African falls, the percentage of each class of meteorite falls observed in Africa is similar to that observed throughout the world, which are, respectively, 84.4% and 86.2% for chondrites, 10,6% and 8.2% for achondrites and 5% and 4.6% for irons. These results confirm the important role that Africa plays in enriching the world collection with observed meteorite falls, in particular, achondrites with the recovery of three Martian meteorites, Nakhla in Egypt, Tissint in Morocco and Zagami in Nigeria. So far, however, as everywhere on Earth, the African collection lacks observed meteorite falls of the lunar type.

The number of observed falls in Africa appears variable in time and space, and has not stopped increasing since 1860, with a high recovery rate during the period 1910-2020 (*i.e.,* 88% of falls recorded) (Table. **3**). Most of these documented falls have been observed and then recorded in three areas of the continent, *i.e.,* North West Africa, East Africa and Southern Africa, in particular in countries that present the combination of the geographical and human factors summarized below.

Table 3. Data of some demographic and geographic factors in African countries. Meteorites data are from "The Meteoritical Bulletin database, 2020". Population and forest cover data are calculated as averages of the 1990-2018 period data from FAO reports.

Countries	Meteorite Falls Number	Meteorite Finds Number	Surface (km²)	Forest Cover (%)	Surface without Forest Cover (km²)	Annual Precipitation (mm)	Population Density (Person/km²)	Total Population (Inhabitants)	Rural Population (Inhabitants)	Rural population (%)	Urban Population (Inhabitants)
Mauritius	1	0	2040	7,84	1880	1824,6	601,1	1226282	534179,389	43,6	692102
Swaziland	1	0	13363	39,51	8083	792,4	86,2	1151257	483050,882	42,0	668207
Malawi	5	1	118480	22,22	92150	1085,5	118,2	14002652	9221446,18	65,9	4781205
Burkina Faso	9	1	274200	25,91	203160	1212,8	53,5	14672320	13198859,2	90,0	1473461
Rwanda	1	0	26338	12,23	23118	1129,9	372,3	9805067	5743237,61	58,6	4061829
Lesotho	1	0	30355	0,46	30215	733,6	64,7	1964136	1506158,13	76,7	457978
Tunisia	5	55	163610	1,23	161600	234,2	62,6	10246355	3587584,06	35,0	6658770
Mali	3	15	124000	10,89	110496,4	317,7	114,5	14192090	9829828,59	69,3	4362261
Morocco	19	1147	710850	6,76	417311	285,4	69,8	31242429	14193975,2	45,4	17048454
Uganda	5	1	236040	18,14	193230	1160,5	131,1	30943925	26618470,8	86,0	4325454
South Africa	22	27	1219912	7,32	1130662	454,6	40,0	48826778	38347042,3	78,5	10479736
Nigeria	16	3	923768	15,06	784618	1125,6	163,1	150637793	91106969,9	60,5	59530823
Zimbabwe	3	1	390580	49,57	196980	641,2	34,8	13596897	9127807,21	67,1	4469090
Kenya	6	1	582650	29,50	410760	658,7	65,3	38037340	21735124,1	57,1	16302216
Tanzania	8	1	945087	41,16	556067	1051	45,4	42886620	31404351	73,2	11482269
Cameroon	3	1	475440	54,85	214680	1534,5	40,9	19424693	9735143,89	50,1	9689549
Niger	9	30	1267000	1,10	1253100	171,4	12,0	15160340	12541928,8	82,7	2618411
Sudan	11	1	2505813	24,98	1879953	419,1	17,3	43306738	28415558,7	65,6	14891179
Mauritania	7	30	1030700	0,32	1027430	91,4	3,3	3374146	2863886,4	84,9	510260
Ghana	1	0	239460	26,96	174910	1169,1	95,9	22954918	11754754,5	51,2	11200164
Angola	3	1	1246700	56,05	547900	970	15,9	19799804	12259084,6	61,9	7540719
Congo Democratic	5	1	2345410	59,92	940100	1519,9	26,3	61652234	48160875,3	78,1	13491358
Ethiopia	5	1	1127127	4,11	1080797	734,7	72,9	82148989	50263306,7	61,2	31885682
Ivory Coast	1	0	322460	22,89	248640	1389,3	59,4	19145190	13791377,1	72,0	5353813
Somalia	2	1	637657	11,87	561937	252,4	14,3	9090922	5674330,31	62,4	3416592
Algeria	8	704	2381740	0,89	2360460	82,6	14,6	34784851	12044792,1	34,6	22740059
Namibia	2	15	825418	9,83	744288	265,9	2,6	2108495	1470021,82	69,7	638473
Central Africa	1	0	622984	37,25	390914	1338	6,8	4249578	1725058,42	40,6	2524520
Zambia	1	0	752614	42,65	431644	888,3	17,5	13135086	8086078,58	61,6	5049008
Madagascar	2	0	587295	20,17	468855	1430,3	33,7	19768968	4474040,91	22,6	15294927
Egypt	2	65	1001450	0,07	1000710	41,8	78,7	78792133	18222041,7	23,1	60570091
Tchad	1	1	1284000	9,95	1156260	330,9	8,6	11072540	9122303,03	82,4	1950237
Lybia	1	1474	1759540	0,20	1756010	47	3,4	5941368	3077850,22	51,8	2863518
Mozambique	0	0	801590	38,26	494940	977,3	28,5	22830132	13451617,3	58,9	9378514
Botswana	0	10	600370	20,90	474920	379,3	3,3	1962216	889599,289	45,3	1072617
Congo Brazaville	0	0	342000	65,01	119650	1619,9	11,2	3820777	2346464,37	61,4	1474313

(Table 3) cont.....

Countries	Meteorite Falls Number	Meteorite Finds Number	Surface (km²)	Forest Cover (%)	Surface without Forest Cover (km²)	Annual Precipitation (mm)	Population Density (Person/km²)	Total Population (Inhabitants)	Rural Population (Inhabitants)	Rural population (%)	Urban Population (Inhabitants)
Gabon	0	0	267667	81,92	48397	1814,5	5,5	1462323	1171234,46	80,1	291089
Senegal	0	0	196190	31,84	133720	712,7	62,4	12240842	4970026,82	40,6	7270816
Erythrea	0	0	121320	13,11	105420	234,6	36,7	4450900	2534431,37	56,9	1916468
Benin	0	0	112620	24,15	85420	1032,3	78,8	8870563	5243510,75	59,1	3627052
Liberia	0	0	111370	31,94	75800	2352,8	32,9	3664930	2835413,15	77,4	829517
Sierra Leone	0	0	71740	15,21	60830	2511,1	76,1	5456157	2625735,74	48,1	2830421
Togo	0	0	56785	9,35	51475	1125,4	106,1	6027587	3839079,75	63,7	2188507
Guinea Bissau	0	0	36120	61,16	14030	1559,3	43,1	1557428	255534,306	16,4	1301894
Equatorial Guinea	0	0	28051	66,24	9471	2176,9	24,3	680407	314157,52	46,2	366249
Burundi	0	0	27830	3,92	26740	1212,6	319,2	8884646	6762766,1	76,1	2121880
Djibouti	0	0	23200	0,26	23140	184	34,5	799573	408775,247	51,1	390798
Gambia	0	0	11300	42,21	6530	612,8	140,3	1585433	1320769,61	83,3	264664
Cap Verde	0	0	4033	22,32	3133	394,3	117,6	474328	180260,922	38,0	294067
Comoros	0	0	2190	3,65	2110	1677,2	301,5	660303	405001,047	61,3	255302
Sao-Tome	0	0	1001	26,97	731	2062,6	162,6	162811	121082,943	74,4	41728
Seychelles	0	0	455	65,93	155	1674,2	188,2	85633	49840,399	58,2	35792

A large population is evenly distributed in the country. The high density of the population in a given place favors the observation and then the recovery of meteorites. The decrease of the African population in the 19th century would explain the low rate of recorded meteorite falls during this period. As well, the number of falls recorded can be related to a wide distribution of population as it occurs in Nigeria and South Africa, while the sparse and uneven population distribution in many other countries makes it difficult to observe meteorite falls. As a result, the falls/founds ratio in Africa (1:60) is much lower than that of other continents or large countries of similar size (*e.g.*, USA 1:13). Furthermore, in Africa, the rate of observed falls appears to be in good agreement with the high proportion (60%) of rural population, as well as with their lifestyle that privileges the observation of meteors and the collection of meteorites. The total mass of recovery of observed meteoroid fall fragments, besides depending on the mass of the meteoroid entering the Earth atmosphere, is also related to the mobility of people, *i.e.,* the more mobile the people are, the more fragments are collected. Indeed, the largest masses were collected in heavily populated areas.

The importance is given to meteorites. The rate of observed meteorite falls is also related to the degree of interest shown in these extraterrestrial rocks by the population of a country. This interest was born in certain African countries thanks to the educational impact of the colonizers in the field of meteorites, which contributed, during the 20th century, to the increase in the rate of observed falls in the continent. Currently, in some African countries, the work of information on the meteor phenomenon performed within the population by simplifying the concept of meteorites and coaching and accompanying the collectors of meteorites, has enabled the recruitment of inhabitants, particularly in desert regions. In Morocco, smaller and less populated than many other countries in Africa, 11 falls were recorded, totaling the highest mass collected continent-wide (384 kg), thanks to the efforts of nomads and rural population who are interested in this spectacular phenomenon and respect the scientists and help them to recover the fragments of meteorites.

Large countries with a low percentage of forest cover. Generally, few or no falls are recorded in large countries covered with dense rainforests with tall trees over large areas, that makes it difficult to perceive meteors and collect their fragments after the explosion.

Semi-arid to arid climate. Most African countries bordering the two continent deserts have recorded relatively high recovery rates, with 75% of meteorite falls observed during the day thanks to the clear skies that prevail for much of the year. Moreover, the contrast between meteorite fragments and soil and cultivated land and sparse grasslands makes it easier the recovery of the falls in these arid and semi-arid zones.

Access to the fall location. In Africa, 86.1% of observed falls meteorites were recovered less than 50 km from the roads. Thus, a low density of road network implies several meteorites be lost and left in their place of fall. Fortunately, the life style of nomads in regions far from roads often allows to discover otherwise lost meteorites. However, some African countries, like South Africa and Algeria, do not allow ordinary people to search, buy and sell meteorites. In particular, the severe application of the law in South Africa since 1999 and in Algeria since 2004 would explain the stability of the number of falls observed and listed after these years in these countries, and the low masses collected, *i.e.,* between 0.0014 and 7 kg for 86% of South African meteorites.

Thus, the rate of observed meteorites falls in each country of Africa depends on one or more geographical and human factors mentioned above. The average rate of falls remains low in Africa, *i.e.,* a single fall per 1.35 years. Thus, certain measures should be undertaken to increase the rate of observed falls. In particular, the cooperation with ordinary people should be strengthened by proposing and supervising training and supporting programs for the inhabitants of rural areas, especially in countries that benefit of human and geographical conditions favorable to the observation of falls and collection of fallen meteorites. These measures are expected to increase the interest of the local population in the recovery of new fragments of old falls or even new discoveries. According to

Beech (2003), in Canada the more the public interest is promoted, the more meteorite fall events are recorded. Furthermore, "meteorites" should be included in African educational systems, which will surely increase the interest of young people in these unique rocks.

Finally, the installation of all-sky cameras for monitoring meteors with precise orbit determination in some African countries is also expected to increase the number of observed meteorites falls, as is the case of France where 100 cameras scan the French territory (Colas *et al.*, 2015).

CONCLUSION

This statistical study shows that in Africa, the rate of observed meteorite falls is low compared to other regions of the world and that certain human and geographical conditions are necessary to promote the observation of these falls and the collection of fallen meteorites. Although observing the penetration of a meteor into the African atmosphere during a clear sky day and then collecting its fragments on sand or in a meadow near a road is not rare, the discovery of fragments of a meteorite fall in a thick forest, results in difficult even with a well distributed dense population in the area of the fall.

This study also highlights those African countries that have made good progress in the general understanding of these extraterrestrial rocks and the sensitization of the population towards meteorites, and other countries that need to develop and strengthen the programs of support that can contribute to expanding the meteorite heritage of this continent.

REFERENCES

Anders, E. (1971). Meteorites and the early solar system. *Annu. Rev. Astron. Astrophys., 9*(1), 1-34.
[http://dx.doi.org/10.1146/annurev.aa.09.090171.000245]

Beech, M. (2003). Canadian fireball rates and meteorite falls – declining returns? *Meteorites Magazine, 9*, 3-9.

Bevan, A.W.R. (2006). The western australian museum meteorite collection. *Spec. Publ. Geol. Soc. Lond., 256*(1), 305-323.
[http://dx.doi.org/10.1144/GSL.SP.2006.256.01.15]

Bevan, A.W.R. (1992). Australian meteorites. *Rec. Aust. Mus.,* (Suppl. 15), 1-27.

Bland, P.A., Berry, F.J., Pillinger, C.T. (1998). Rapid weathering in Holbrook: An iron-57 Mössbauer spectroscopy study. *Meteorit. Planet. Sci., 33*(1), 127-129.
[http://dx.doi.org/10.1111/j.1945-5100.1998.tb01614.x]

Burke, J.G. (1991). Cosmic debris: Meteorites in History. University of California Press. 455.

Chennaoui, A.H., Jambon, A., Bourot, D.M. (2009). Tamdakht Meteorite: The Last Moroccan Fall. *Meteoritics & Planetary Science, 44*.

Clarke, J.R., Howard, P., Timothy, J.M. (2006). Meteorites and the Smithsonian Institution ROY S.The History of Meteoritics and Key Meteorite Collections: Fireballs, Falls and Finds. The Geological Society of

London, Special Publications 256, 1-13.

Colas, F., Zanda, B., Bouley, S., Vaubaillon, J., Marmo, C., Audureau, Y., Kwon, M.K., Rault, J.L., Caminade, S., Vernazza, P., Gattacceca, J., Birlan, M., Maquet, L., Egal, A., Rotaru, M., Gruson Daniel, Y., Birnbaum, C., Cochard, F., Thizy, O. (2015). FRIPON, the French fireball network. *Proceedings International Meteor Conference IMO34,* 34-38.

Cooke, B., Brown, P., Blaauw, R., Kingery, A., Moser, D. (2015). Bright Fireball Characterization and Modeling. *In 2015 AlaSim International Conference (No. M15-4538).*

Dodd, R.T. (1986). Thunderstones and shooting stars: the meaning of meteorites. Harvard University Press, Cambridge, Massachusetts and London, England, 196.
[http://dx.doi.org/10.4159/harvard.9780674284975]

FAO. (2002). Evaluation des ressources forestières mondiales. Rome, 140.

Gould, T. (2003). The southern african meteorite recovery program. *Monthly Notes of the Astronomical Society of Southern Africa,* *62*, 124-130.

Gounelle, M. (2009). Les météorites.. Presses universitaires de France. 128.
[http://dx.doi.org/10.3917/puf.goune.2009.01]

Graham, A.L., Bevan, A.W.R., Hutchison, R. (1985). *The Catalogue of Meteorites.* Trustees of the British Museum (National History).(Fourth ed..). London.:

Halliday, I. (2001). The present-day flux of meteorites to the Earth. In: Peuker-Ehrenbrink, B., Schmitz, B., (Eds.), *Accretion of Extraterrestrial Matter Throughout Earth's History.* (pp. 305-318). New York: Kluwer.
[http://dx.doi.org/10.1007/978-1-4419-8694-8_16]

Heide, F. (1957). *Klein Meteoritenkunde.* Springer.(Second ed..). Verlage, Berlin (Translated as Meteorites, 1964, University of Chicago Press):
[http://dx.doi.org/10.1007/978-3-662-12147-4]

Hughes, D. (1992). The meteorite flux. *Space Sci. Rev., 61*(3-4), 275-299.
[http://dx.doi.org/10.1007/BF00222309]

Ibhi, A. (2013). New Mars meteorite fall in Morocco: final strewn field. International Letters of Chemistry. *Physics and Astronomy, 11*, 20-25.

Ibhi, A. (2014). Tighert: A new eucrite meteorite fall from morocco. *International Meteor conference, 33*, 1-3.

Ibhi A. (2016). Météorites : Perles du Désert Marocain. Souss Impression, 241.

Ibhi A. Nachit, H., Abia El, H. (2013). Tissint meteorite: new mars meteorite fall in morocco. *J. Mater. Environ. Sci., 4*, 293-298.

Ibhi, A. (2013). Meteors and meteorite falls in Morocco. International Letters of Chemistry. *Physics and Astronomy, 12*, 28-35.

Ingram, K.T., Roncoli, M.C., Kirshen, P.H. (2002). Opportunities and constraints for farmers of west Africa to use seasonal precipitation forecasts with Burkina Faso as a case study. *Agric. Syst., 74*(3), 331-349.
[http://dx.doi.org/10.1016/S0308-521X(02)00044-6]

Koschny, D., Soja, R.H., Engrand, C., Flynn, G.J., Lasue, J., Levasseur-Regourd, A-C., Malaspina, D., Nakamura, T., Poppe, A.R., Sterken, V.J., Trigo-Rodríguez, J.M. (2019). Interplanetary Dust, Meteoroids, Meteors and Meteorites. *Space Sci. Rev., 215*(4), 34.
[http://dx.doi.org/10.1007/s11214-019-0597-7]

Kress, M. (2001). Collecting cosmic dust. *Mercury Magazine Contents, 30*, 24-31.

Krot, A.N., Keil, K., Scott, E.R.D. (2013). Classification of meteorites and their genetic relationships. Meteorites. *Comets and Planets, 1*, 1-63.
[http://dx.doi.org/10.1016/B978-0-08-095975-7.00102-9]

Lin, Y., El Goresy, A., Hu, S., Zhang, J., Gillet, P., Xu, Y., Hao, J., Miyahara, M., Ouyang, Z., Ohtani, E., Xu, L., Yang, W., Feng, L., Zhao, X., Yang, J., Ozawa, S. (2014). NanoSIMS analysis of organic carbon from the Tissint Martian meteorite: Evidence for the past existence of subsurface organic-bearing fluids on Mars. *Meteorit. Planet. Sci., 49*(12), 2201-2218.
[http://dx.doi.org/10.1111/maps.12389]

Diop-Maes, L-M., Diop, M. (1985). Essai d'évaluation de la population de l'Afrique Noire aux XVe et XVIe siècles. *Population (Paris), 40*(6), 855-884.
[http://dx.doi.org/10.2307/1532781]

Mason, B. (1963). Olivine composition in chondrites. *Geochim. Cosmochim. Acta, 27*(10), 1011-1023.
[http://dx.doi.org/10.1016/0016-7037(63)90062-0]

McCubbin, F.M., Shearer, C.K., Burger, P.V., Hauri, E.H., Wang, J., Elardo, S.M., Papike, J.J. (2014). Volatile abundances of coexisting merrillite and apatite in the martian meteorite Shergotty: Implications for merrillite in hydrous magmas. *Am. Mineral., 99*(7), 1347-1354.
[http://dx.doi.org/10.2138/am.2014.4782]

Meteoritical bulletin database (2020). The meteoritical society. Available at: http://www.lpi.usra.edu/meteor

Molau, S. (2013). Meteor showers identified from one million video meteors In: Gyssens, M., Roggemans, P., Zoladek, P., (Eds.), *Proceeding of the International Meteor Conference IMO 32, 26-38.

Notkin G. (2011). Meteorite Hunting: How to find treasure from space. Aerolite Meteorites, ed. Stanegate Press, 84.

Rasmussen, K.L. (1991). Historical accretionary events from 700 BC to AD 1850-A 1050-year periodicity. *Quarterly Journal of the Royal Astronomical Society, 31*, 95-108.

Rubin, A.E., Grossman, J.N. (2010). Meteorite and meteoroid: New comprehensive definitions. *Meteorit. Planet. Sci., 45*, 114-122.

Scott, E.R.D. (1977). Pallasites—metal composition, classification and relationships with iron meteorites. *Geochim. Cosmochim. Acta, 41*(3), 349-360.
[http://dx.doi.org/10.1016/0016-7037(77)90262-9]

SWAC, (2015). Population density in the Sahara. Maps and Facts 14, SWAC Publishing, Paris.

Treiman, A.H. (2003). The Nakhla Martian meteorite is a cumulate igneous rock: Comment on Varela M.E., Kurat G., Clocchiatti R., (2001). Glass-bearing inclusions in Nakhla (SNC meteorite) augite: heterogeneously trapped phases. *Mineral. Petrol., 77*, 271-277.
[http://dx.doi.org/10.1007/s007100300000]

Treiman, A.H. (1993). The parent magma of the Nakhla (SNC) meteorite, inferred from magmatic inclusions. *Geochim. Cosmochim. Acta, 57*(19), 4753-4767.
[http://dx.doi.org/10.1016/0016-7037(93)90198-6]

Tschermak G., (1885). Die mikroskopische Beschaffenheif der Meteorifen. Translated by Wood. J.A. and Wood. E.M. and reprinted in 1964 in smithsonian contributions to astrophysics 4, 138-239.

Wasson J.T., & Wetherill G.W., (1979). Dynamical, chemical and isotopic evidence regarding the formation locations of asteroids and meteorites, in Gehrels T. (Eds.), Asteroids. Univeristy of Arizona Press, Tucson, 926-974.

Weisberg M.K., McCoy T.J., & Krot A.N., (2006). Systematics and evolution of meteorite classification, in Lauretta D.S., McSween H.Y. (Eds.), Meteorites and the Early Solar System II, University of Arizona Press, Tucson, 19-52.
[http://dx.doi.org/10.2307/j.ctv1v7zdmm.8]

Wetherill, G.W. (1968). Stone Meteorites: Time of Fall and Origin. *159, 7*, 9-82.

Wickman, F.E., Palmer, C.D. (1979). *Proc. Indiana Acad. Sci., 88*, 247-272.

CHAPTER 3

Meteorite Finds in Africa

Lahcen Ouknine[1,2,*], **Giorgio S. Senesi**[3], **Fouad Khiri**[1,2], **Abderrahmane Ibhi**[1,2] and **Mohamed Th. S. Heikal**[4]

[1] *Geoheritage and Geomaterials Laboratory, Ibn Zohr University, Agadir, Morocco*

[2] *University Museum of Meteorites, Ibn Zohr University, Agadir, Morocco*

[3] *CNR, Institute for Plasma Science and Technology (ISTP), Bari seat, 70126 Bari, Italy*

[4] *Geology Department, Faculty of Science, Tanta University, Tanta, Egypt*

Abstract: Africa is a favorable site for meteorite recovery, with a total number of recoveries amounting to more than 1/6 of all meteorites recovered from the entire world. This work deals with the classification of meteorite finds in Africa, the distribution of their masses, and their alteration/weathering grades as affected by various factors. The African meteorite population includes an abundance of stony meteorites with a high percentage of the world collection of rare meteorites, *i.e.*, Martian meteorites (62%), Ureilites (51%), Rumuruti (59%), Lunar (47%), and HED (46%). Furthermore, an important increase in achondrite meteorites finds occurred in the last two decades, compared to the Australian and Antarctic collections. The mass distribution of the African meteorite population shows that most recoveries (72%) have masses bigger than 100 g with peaks of about 1 kg, compared to about 0.1 kg for the Australian collection and 0.01 kg for the Antarctic finds. The distribution of weathering grades (W) shows the predominance of W1 (32%) and W2 (34%), which proves a better preservation of meteorites in this continent. The factors influencing the mechanism and rate of alteration of African finds include climate as the main factor, the mass, the terrestrial age, and the initial porosity of the sample.

Keywords: Africa, Classification, Human and natural factors, Mass distribution, Meteorite finds, Weathering factors.

INTRODUCTION

The majority of meteorites collected on Earth originate from the "cold" Antarctica deserts, where they are embedded within the ice (Whillans *et al.*, 1983; Corti *et al.* 2003), and "hot" deserts (Bland *et al.*, 2000), such as the Sahara in North West Africa (NWA) (Bischoff and Geiger, 1995; Schlüter *et al.*, 2002; Ibhi, 2014,

* **Corresponding author Lahcen Ouknine:** Geoheritage and Geomaterials Laboratory, Ibn Zohr University, Agadir, Morocco & University Museum of Meteorites, Ibn Zohr University, Agadir, Morocco; E-mail: lahcen.ouknine@edu.uiz.ac.ma

2016; Khiri *et al.*, 2017), the desert of Oman (Al-Kathiri *et al.*, 2005; Hofmann & *et al.*, 2018), the Atacama desert in Chile (Muñoz *et al.*, 2007; Valenzuela & Benado, 2018), and the arid regions of Australia (Nullarbor) (Bevan & Bindon, 1996. Jull *et al.*, 2010) and United States of America (Roosevelt County) (Bland *et al.*, 2000; Jull *et al.*, 2010). In particular, the African meteorite collection represents the second largest one after that of Antarctica and includes meteorites of great scientific value.

Globally, only 10 countries/regions can document more than 300 meteorite finds within their borders. These include five countries/regions in Africa, *i.e.*, NWA with 9387 findings, Sahara with 476, Libya with 1504, Morocco with 1741 and Algeria with 609. Until January 1, 2021, a total of 12511 meteorite discoveries of African origin are listed in the Meteoritical Bulletin Database of the Meteoritical Society (www.lpi.usra.edu/meteor/metbull.php). Khiri *et al.* (2017) reported that almost all African countries bordering deserts have a relatively high meteorite recovery rate, which could simply be a consequence of their easy viewing of the desert surface.

The first part of this chapter aims to review the spatiotemporal distribution of meteorite finds in Africa and evaluate the human and natural factors that have favored their discovery. The second part is devoted to analyze the typology and classify the meteorites collected and their mass distribution. The third part deals with the alteration/weathering processes to which African meteorites have been subjected and the various influencing factors. The chapter ends with a brief summary and conclusion. Along with the entire review, the African meteorite features are compared with those of other meteorite populations, especially from Antarctica and Australia.

FACTORS INFLUENCING METEORITES FINDS IN AFRICA

Historical and Cultural Factors

The rate of African meteorite finds was low in the 18th and 19th centuries, with 2 and 10 meteorites collected, respectively, whereas the number of finds increased in the 20th century, especially towards its end and in the North of the continent with the collection of 2417 samples (Fig. **1**). In the 21st century, the meteorite recovery rate almost quadrupled between 2000 and 2016, with 9888 meteorites recorded.

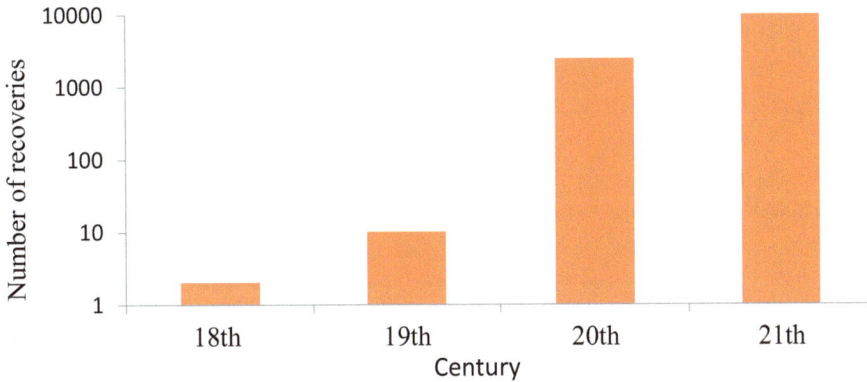

Fig. (1). Distribution by the century of meteorite finds in Africa.

The low rate of finds in the 18th and 19th centuries can be explained by the lack of knowledge and culture on meteorites. In antiquity, in many African countries, the passing of meteors was related to superstitious special events such as the death of someone or the passing of the devil (Khiri *et al.*, 2017). Historians have described that many ancient African civilizations regarded both meteorites and the site of its fall as sacred. Often meteorites have been used as objects of worship or to produce handicrafts such as jewelry, weapons and practical items (Johnson *et al.*, 2015; Comelli *et al.*, 2016). The colonization of African countries in the 20th century modified the previous beliefs of natives about meteorites. Dodd (1986) and Bevan (1992) noted that historical and educational factors also contributed to meteorite recovery in Australia.

The period between the end of the 20th and the beginning of the 21st century (1986-2020) was characterized by the collection of hundreds or even thousands of meteorites by systematic expeditions organized particularly in the hot deserts of northern Africa, *i.e.*, in Morocco, Algeria, and Libya. This was promoted by both (i) the major meteorite discoveries recorded in hot deserts around the world (Dhofar in Oman, Nullarbor in Australia, Atacama in Chile, and Roosevelt County in USA) as the result of systematic missions initiated at the start of the '70s, and (ii) the interest in planetary sciences with the birth of research teams focusing on meteorites in Morocco, Algeria and South Africa.

In some African countries, meteorites were often supplied by nomads and locals who swept the desert in search of meteorites to sell. Beech (2003) suggested that to recover more meteorites, we need both an educated and inspired local population and subject matter scientists in the field. In USA in the 30's, meteorite search missions were organized for the first time with the involvement of farmers

(Nininger, 1972). However, according to Clarke *et al.* (2006), until 1984, in USA, citizens were passionate about the phenomena of falling meteorites, but not consider them high-priority objects to acquire. Rasmussen (1991) reported that the distribution of the human population affects the probability of recognition and recovery of meteorites. The current state of the spatial distribution of meteorite finds in Africa as a function of population density is shown in Fig. (**2**).

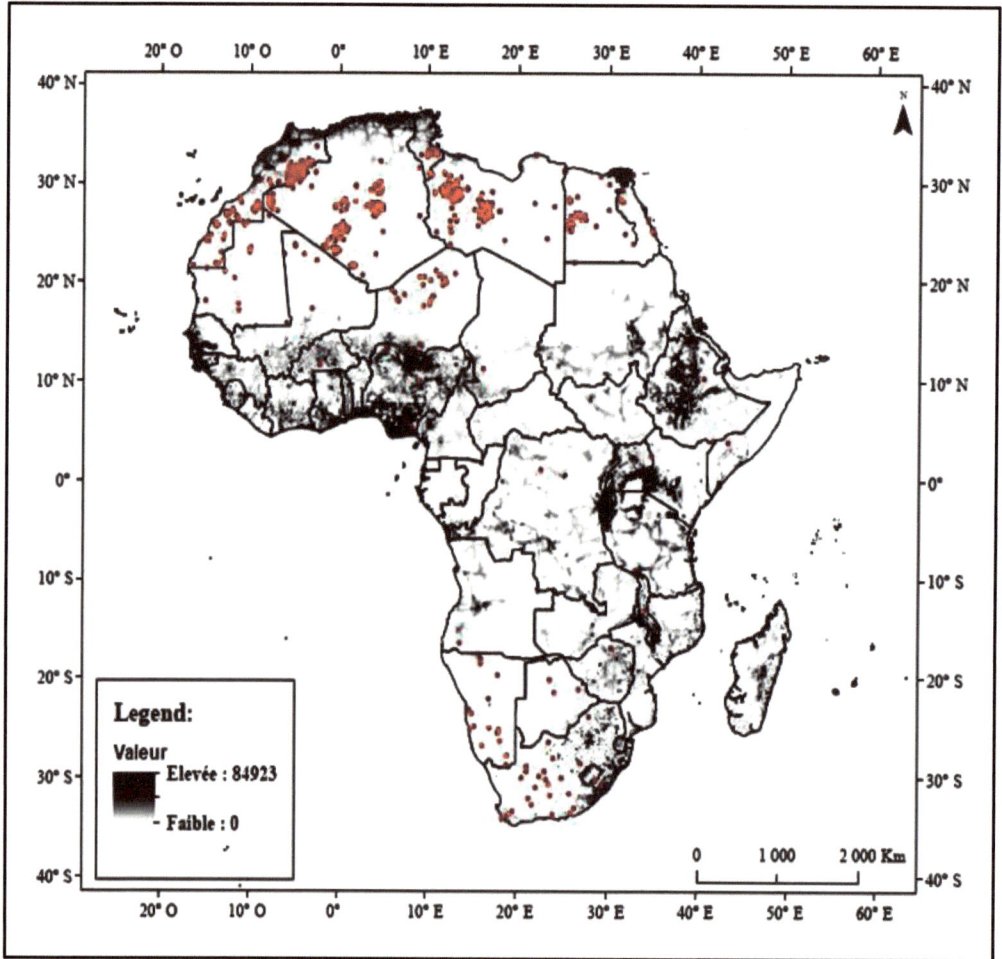

Fig. (2). Map showing the distribution of different meteorite finds in Africa as a function of population density.

Apparently, most of finds in Africa are concentrated in areas with a desert climate, *i.e.*, (i) in the North, mainly in the semi-arid to arid regions of Morocco, Algeria, Tunisia, Libya, Egypt, and in the northern margin of the Sahara Desert

(Mauritania, Mali, Niger), in the South of the continent, mainly in southwest particularly in the Namib and Kalahari deserts. Despite their low population density (less than five people per km^2 in Sahara), the desert regions record significant meteorite discoveries, which yields a negative correlation between the number of finds and population density. However, in some countries, such as Libya and Egypt, the rate of discovery is high even with an unevenly distributed population, whereas in many countries in the middle of the continent, such as Nigeria, Ethiopia, Tanzania, *etc.*, the rate of meteorite discoveries is low despite the high population density.

Although in Africa, a positive correlation exists between the distribution of observed meteorite falls and population density (Khiri *et al.*, 2016), the rural, urban and total populations show no correlation with the number of meteorite finds (correlation coefficients of, respectively, 0.03, 0.028 and 0.10). This means that the uniform distribution of population density does not facilitate the discovery of meteorites in Africa. Thus, the distribution of meteorite finds in the continent can be explained by the interest in space exploration and the increase in research teams interested in meteorites, especially since the beginning of the 21st century. Furthermore, nomads, collectors and meteorite hunters have developed a real passion for meteorite recovery.

To preserve the Moroccan meteorological heritage, in the year 2016, the first meteorite museum in Africa (Fig. **3**), the "University Meteorite Museum" in Agadir, which exhibits more than 100 meteorites of different types (Ibhi, 2016), was created. Furthermore, several initiatives have been implemented at the continental level to strengthen scientific research on meteorites and promote the sciences of astronomy. In particular, besides the South African National Space Agency, another agency was created in Egypt, the Egyptian Space Agency, and 8 astronomical observatories were installed in 6 countries to study astronomical phenomena, especially the detection of meteors, and facilitate the recovery of fallen meteorites. Moreover, as part of the Desert Fireball Network (DFN) program (Bland, 2016), portable cameras have been installed in Morocco to detect falling meteorites and thus accelerate their recoveries. In 2017, researchers from some African countries launched an initiative to promote planetary and space science in the continent called the "African Initiative for Planetary and Space Sciences" (Baratoux *et al.*, 2017). Finally, several research units have been created in various countries, *i.e.*, Morocco, Algeria, Egypt, and South Africa, to promote research in planetary sciences and specialized associations, *e.g.*, the Moroccan Spacenauts, organize sharing activities aimed at popularizing the planetary sciences to the public.

Fig. (3). Visitors of different social groups at the Agadir University Meteorites Museum.

Natural Factors: Climate and Plant Coverage

Meteorite finds in Africa are mainly concentrated in the hot deserts of the North and south-west of the continent (Namib and Kalahari deserts). These two regions are characterized by a semi-arid to arid climate in which rainfall is too low to support vegetation (25 to 200 mm per year), and most water is lost through evapotranspiration (Laity, 2009). For example, in the Dar al Gani region of Libya, where a meteorite dense collection area (DCA) is located, the average annual precipitation is 10 to 20 mm, and rainwater is quickly leached into the underground karst system (Schlüter *et al.*, 2002). In contrast, the number of meteorites collected in tropical African countries is low, and some countries, *e.g.*, Côte d'Ivoire, Ghana and Angola, have never recorded any meteorite as in other tropical regions in the world, *e.g.*, Cambodia (0 finds, 3 falls), Thailand (1 find, 3 falls), Indonesia (3 finds, 17 falls) and Bolivia (2 discoveries, 1 fall).

The distribution of various meteorite finds in Africa in relation to the density of land use Fig. (**4**) shows that meteorite finds are mainly distributed in the North of the continent in semi-arid and arid zones that are characterized by sparse vegetation consisting of thorny herbaceous plants, shrubs and grasses resistant to drought. In the southwest, meteorites are collected in areas with closed-to-open meadows composed of herbaceous plants and thermophilic perennial shrubs resistant to drought. In the countries of North Africa, an average correlation exists between the percentage of forest cover, which expresses the proportion of forested areas in relation to the total area of a country, and the number of findings. Differently, in the tropical region, the number of meteorite finds is very low (15 finds). Furthermore, a negative correlation exists between the number of samples collected and the forest cover ratio.

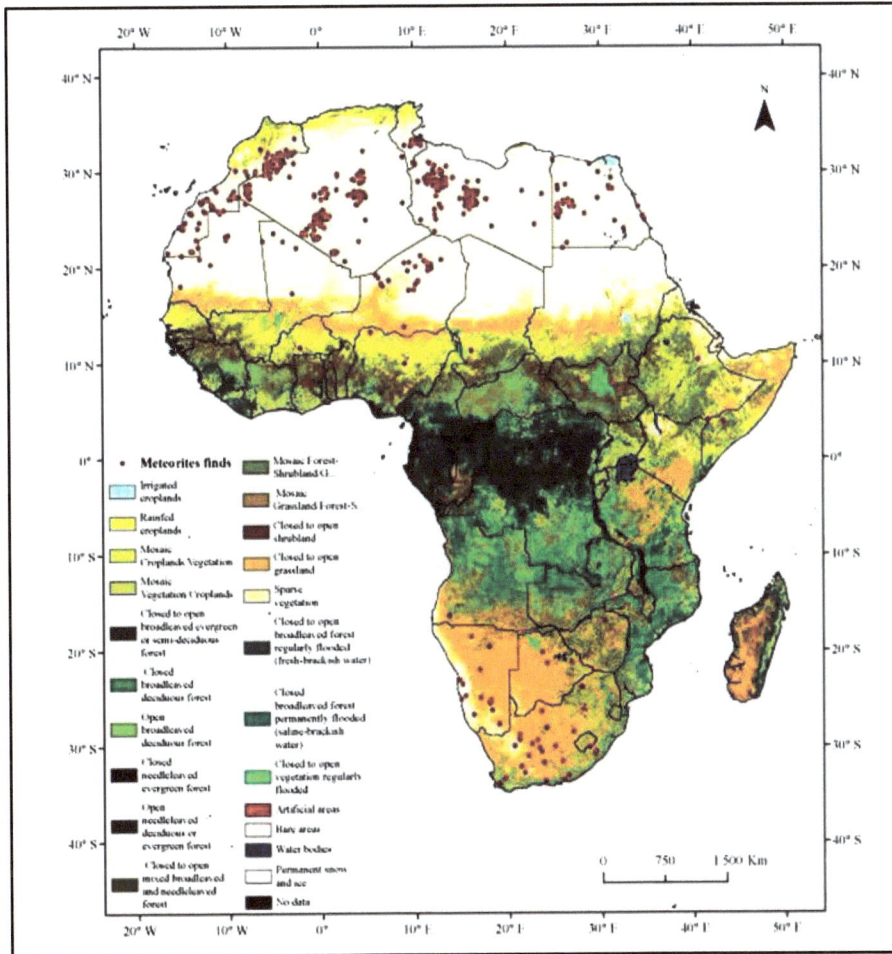

Fig. (4). Distribution of various meteorite finds in Africa in relation to the density of land use.

The distribution of the 4471 African meteorite finds recovered as a function of the well-known month of discovery Fig. (**5**) shows that the maximum number, *i.e.*, 1225 samples collected, was recorded in April and the minimum one, *i.e.*, 149 and 167, in July and August, respectively. Most of the finds are confirmed to originate from semi-arid to arid areas of North Africa, where temperatures can reach 55 ° C in summer and drop to -10 ° C overnight in winter. The largest meteorite finds take place in April, as this month is characterized by a moderate temperature agreeable to a meteorite.

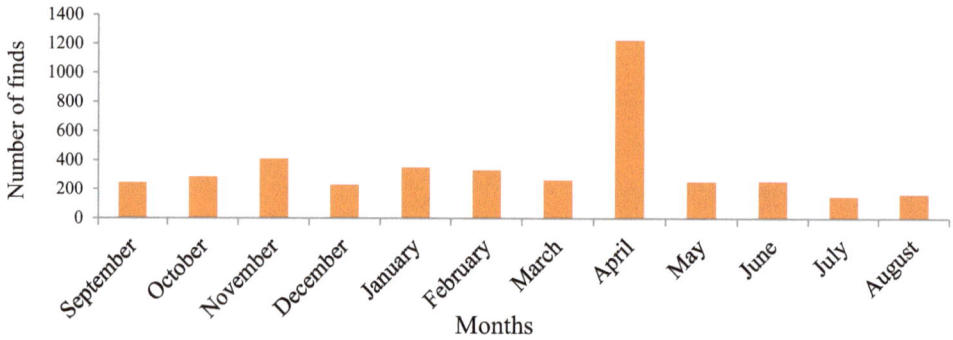

Fig. (5). Variation in the number of meteorites finds in Africa per month of recovery.

The number of meteorites collected in dry (spring and summer) and wet (autumn and winter) seasons Fig. (**6**) indicates that the former one is greater (56%) than the latter one. This confirms that the wet season is suitable for meteor hunters, nomads and scientists for collecting meteorites in semi-arid and arid areas of northern deserts, Namib and Kalahari, which lack shelters (trees and shrubs are rare) to protect from the strong heat. Thus, scientific expeditions for prospecting the Sahara Desert (DCA Tanezrouft, DCA Dar el Gani) in search of meteorites were organized during this period (Bischoff and Geiger, 1995; Schlüter *et al.*, 2002).

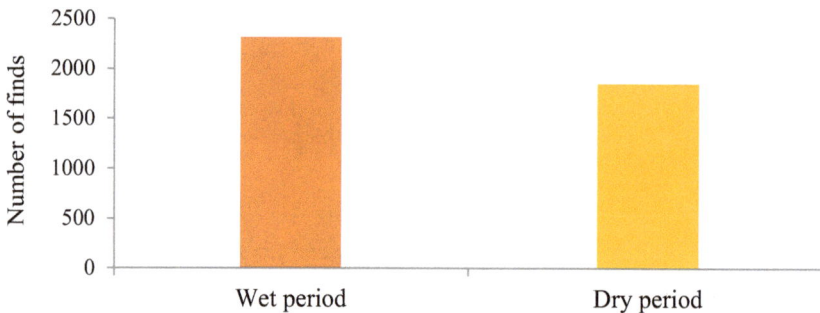

Fig. (6). Comparison of meteorite find rates in Africa between the dry season (spring and summer) and wet season (autumn and winter).

A more detailed evaluation of the number of meteorites finds by season Fig. (**7**) shows that the number of finds in autumn (931) is very similar to that in winter (937), whereas the finds doubles in spring (1738 samples) and reaches a minimum (568 samples) during summer. As a result, most meteorite researchers prefer the spring season, especially April, which is characterized by climatic conditions very favorable for meteorite recovery.

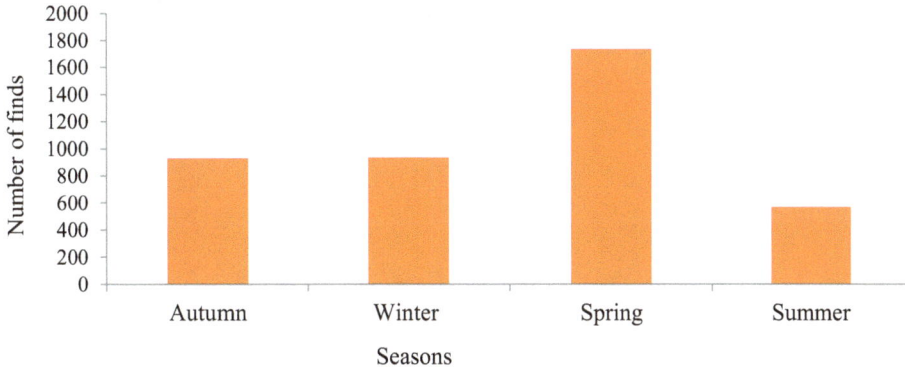

Fig. (7). Variation in the number of meteorites discovered per season in Africa.

In conclusion, the high rate of meteorite finds in Africa does not appear to depend on the total population density or its distribution, but is related to a category of the population interested in these extraterrestrial rocks, *i.e.*, nomads, passionate people, scientists, meteorite hunters, *etc.*, provided with skill appropriate to distinguish extraterrestrial from terrestrial rocks. The search for meteorites in Africa is carried out in hot deserts with sparse vegetation cover, during humid periods (autumn and spring), especially in April.

AFRICAN METEORITE FINDS: CLASSES AND MASS DISTRIBUTION

Relative Abundance by Class and Type

Before 1794, the year in which Ernst Chladni assessed the extra-terrestrial origin of meteorites, two iron meteorites were discovered in Africa, the first one in 1716 in Mali (named Siratek) and the second one in 1793 in South Africa (named Cape of Good Hope). During the past three centuries, 12511 meteorite finds of different types were recorded, totaling a mass of 14815 kg. Note that in the following text, based on the database of the Meteoritical Society, some meteorites are classified with the same name and type, especially in DCAs (Larouci *et al.*, 2014), which suggests that some of these samples could be matched to another officially named sample in the event of overwhelming evidence. Thus, this matching problem can bias the total number and mass in a small meteorite population, but this is not the case in this study in which all meteorite finds from the database have been considered as individuals.

The total number of finds by class (stones, stony-irons, and irons) recovered each decade between 1710 and 2020 in Africa is shown in Fig. (**8**).

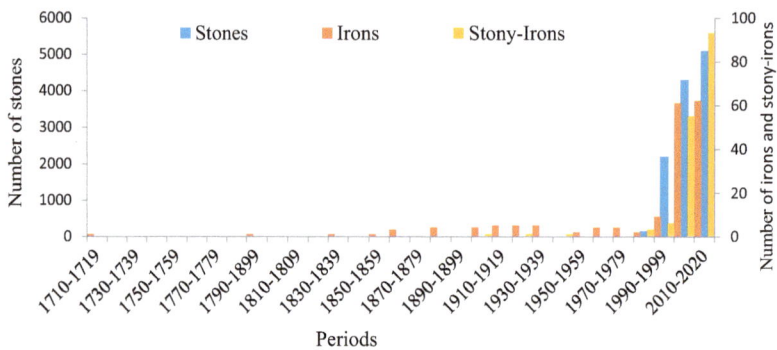

Fig. (8). Meteorite finds (stones, stony-irons, and irons) in Africa between 1710 and 2020. Data extracted from the Meteoritical Bulletin Database.

The recoveries in Africa were strongly influenced by the large amount of meteorites found from the year 1976 in Antarctica (Cassidy and Rancitelli, 1982), in Australia (Jull *et al.*, 2010; Bevan, 1992) and in Chile (Gattacecca *et al.*, 2011; Hutzler *et al.*, 2016). Thus, an extensive dedicated search started in 1986, which led to several hundreds of stony meteorites recovered in the DCAs of Reg el Açfer and Tanezrouft in Algeria, Dar al Gani and Hamada El Hamra in Libya and Sahara Desert (Bischoff and Geiger, 1995). Noteworthy at the beginning of the 21st century (2000-2020), the number of finds increased sharply with the collection of thousands of stony meteorites (9394) and almost all (90%) stony-irons.

In January 2021, the recoveries in the African continent represented more than 19.5% of all meteorite finds recovered worldwide. Africa represents a prolific place for meteorite recovery, with an average concentration estimated to be 0.32 10^{-3} samples per km^2, more than Australia (0.01 x 10^{-3} samples per km^2). The main favorable features in Africa that permit the preservation and accumulation of meteorites and their search are: (i) the favorable geomorphologic features of the landscape, *i.e.*, sand surfaces (Regs) and microdunes (Ergs) (Righter *et al.*, 2014); (ii) the surface stability of relatively young find sites (Bland *et al.*, 2000); (iii) the importance given to extraterrestrial rocks by local people and the increased systematic search after the large discoveries in Antarctica since 1976; and (iv) the liberal legal rules about the collection of meteorites in most African countries.

The relative abundance of classes, types and groups of meteorite finds in Africa, Australia and Antarctica is shown in Table **1**. Apparently, the African meteorite population includes all types of meteorites known until now, with stony meteorites being the most abundant (97.23%) and subdivided into 77% chondrites and about 20.35% achondrites, whereas stony-irons and irons represent only

1.32% and 1.45% respectively of all finds. Chondritic meteorites include 8212 ordinary chondrites subdivided into 3308 of H type, 3494 of L type and 1410 of LL type, and a high number of rare types, *i.e.*, 161 rumuruti, 1038 carbonaceous, 169 Enstatite and 39 ungrouped chondrites. The African achondrite collection comprises almost half (309 samples) of Lunar meteorites worldwide, including the Dar al Gani 262 that was discovered in the Libyan Sahara (Bischoff and Weber, 1997), and more than half (227) of world Martian meteorites. African achondrites also include other rare types, *i.e.*, 150 primitive achondrites, 26 enstatites, 20 aubrites, 1358 HED, 29 angrites, 349 ureilites and 78 ungrouped achondrites. Furthermore, the African finds feature a higher frequency of achondrites (20.35%) compared to the Australian and Antarctic finds (4.74% and 3.32%, respectively.

Table 1. Relative abundance (in %) of different classes, types and groups of African, Australian, and Antarctic meteorite finds. Data extracted from Meteoritical Bulletin Database (January 2021).

Types				Total Number	Abundance (%)		
					Africa	Australia	Antarctica
Stony meteorites	Chondrites	Carbonaceous		1038	8.30	82.45	96.02
		Ordinary	L	3494	27.93		
			LL	1410	11.27		
			H	3308	26.44	76.98	
		Rumuruti		161	1.29		
		Enstatite Chondrite		169	1.35		
		Ungrouped		39	0.31		
	Achondrites	Primitive achondrites		150	1.20	4.74	3.32
		Enstatites		26	0.21		
		Aubrites		20	0.16		
		Angrites		29	0.23		
		HED		1358	10.85		
		Ureilites		349	2.79	20.35	
		Lunar meteorites		309	2.47		
		Martian meteorites		227	1.81		
		Ungrouped		78	0.62		
Stony- Iron meteorites		Pallasites		37	0.30	1.30	0.23
		Mesosiderites		128	1.02	1.32	
Iron meteorites		Iron-Ixx		181	1.45	1.45 / 11.51	0.43
Total				12511	100	100 / 100	100

During the last two decades, the proportion of achondrite finds in Africa has increased markedly with respect to an apparent decrease of chondrite finds (Table 2), especially ordinary chondrites, whereas the populations of Australia and Antarctica have shown an enrichment of both find types. The decreased number of ordinary chondrites might be caused by meteorites collected by locals in Northwest Africa deserts, but remained unclassified and thus not included in the Meteoritical Bulletin Database (Corrigan *et al.* 2014). In particular, ordinary chondrites are weathered faster than metal-poor achondrites because of volume expansion and consequent fracturing associated with metal oxidation (Al-Kathiri *et al.*, 2005). Differently, an important increase was noticed recently for the carbonaceous chondrites and achondrites finds in Africa, although these types of meteorites are very difficult to spot on the ground. This increase can be explained by the experience gained by meteorite hunters in detecting them easier and by the high prices gained in the collector market. Another hypothesis (Dennison *et al.*, 1986; Gattacceca *et al.*, 2011) is that the flux of meteorites changed qualitatively over the falls time scale of a few hundred years. Moreover, the rarest African meteorites represent a high percentage of the world collection, *i.e.*, 62% of Martian meteorites, 51% of Ureilites, 59% of Rumuruti, 47% of Lunar and 46% of HED. The abundance of rare meteorite types could be ascribed to their preservation and accumulation over time in the semi-arid and arid regions of Morocco, Algeria and Libya.

Table 2. Evolution of frequencies (%) of meteorite group finds in Africa from 1998 to 2017. Data extracted from Meteoritical Bulletin Database.

Types	Chondrite	Achondrites	Stony-Irons	Irons	Total Number
Abundance (%) in 1998	94.73	1.99	0.59	2.69	1861
Abundance (%) in 2017	76.98	20.35	1.32	1.45	12511

The African population finds comprise rare stony-iron meteorites represented by 37 pallasites and 128 mesosiderites, and a few iron meteorites, *i.e.*, 181 accounting for 1.45%. Most of these irons were found by geologists and some collectors during the period of intense mining activity spanning from the 19th to the 20th century (*e.g.*, since 1860 in South Africa) (Marchand, 1996). The rarity of the iron finds also observed in Oman (Zurfluh *et al.*, 2013) could be mostly ascribed to the long human occupation of areas such as the Sahara and Atacama in Chile, which implied their collection in pre-historical and historical times to be used as a source of iron (Hützler *et al.*, 2016).

Mass Distribution

The mass of collected meteorite finds in Africa ranges from 0.48 g for the Martian meteorite NWA 8116 up to 60 tons recorded for the Namibian iron meteorite "Hoba". The distribution of African meteorite finds as a function of mass Fig. (**9**) shows that 54% have masses between 0.1 and 1 kg, and that high-mass meteorites are rare relative to small ones, with only six iron meteorites with a gigantic mass exceeding 1 ton, *i.e.*, "Hoba" (60 tons) and "Gibeon" (26 tons) in Namibia and "Mbosi" (16 tons) in Tanzania. Large stony meteorites have been recovered rarely, with the largest single mass recorded to date (3 tons) for the Al Haggounia 002 chondrite H4 type discovered in the southwest of Morocco.

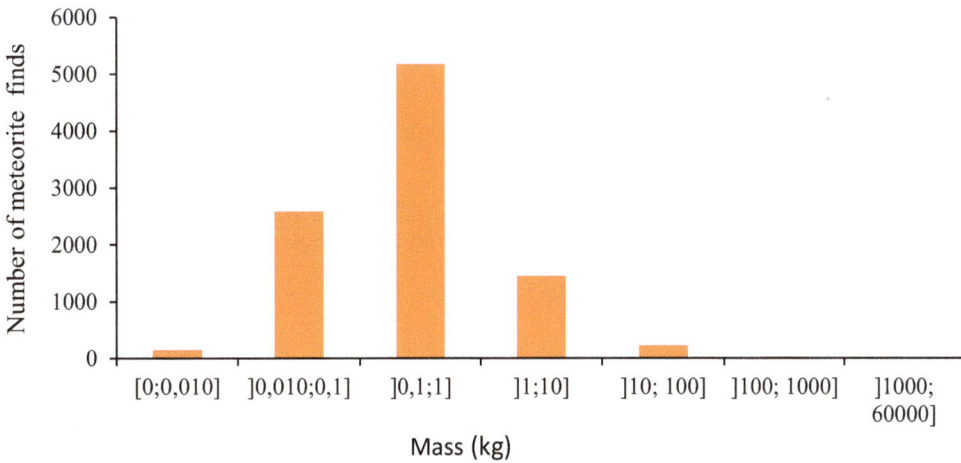

Fig. (9). Masses distribution frequency of meteorite finds in Africa. Data extracted from Meteoritical Bulletin Database.

A survey of meteorite falls on Earth during the last 50,000 years showed that the total mass flux of meteorites of 10 g to 1 kg mass was 2900-7300 kg/year, which represents 15% of all world finds, whereas one quarter (28%) has a mass of less than 100 g (Bland, 2001). However, 80% of African meteorite finds have masses between 10 g and 1 kg, which may be due to the interest in this continent for the recovery of meteorites, the discovery of rare samples such as Martian, Lunar and others during the latter two decades, and a more detailed meteorite exploration by private collectors and mineral dealers in search of meteorites mainly on foot or by car, especially in the desert areas of Morocco, Algeria, and Libya.

The frequency distribution of meteorite finds masses in Africa, Australia and Antarctica Fig. (**10**) shows that the mass of Antarctic meteorites peaks at about 10 g, compared to about 100 g of the Australian collection and about 1000 g of the

African one. However, the Australian population features many smaller meteorites finds with masses between 10 to 50 g (Bevan *et al.*, 1998), whereas the African and Australian mass frequencies decrease low masses (10 g). This difference may be ascribed to the favorable conditions of collecting in Antarctica, as the smaller meteorites are relatively easy to locate due to the generally white to the blue background and, locally, to the low number of admixed terrestrial rocks (Zolensky *et al.*, 2006), so that even small meteorites in a search area can be found (Korotev, 2017). Furthermore, for chondrites, it is impossible to determine in the field if two small stones are paired, so each of them is given a different number (Korotev, 2017). Differently, in Africa and Australia, a 10-g meteorite is practically invisible on the soil, especially if covered by vegetation, thus it should be at least about 10 cm in diameter, *i.e.*, about 2 kg in mass, to be efficiently spotted (Huss, 1991). In conclusion, the differences in the distribution of meteorite masses among the three populations can be mostly explained by the differences in surface features, *i.e.*, the accessibility of location areas and the collecting methods, *i.e.*, on foot or by car.

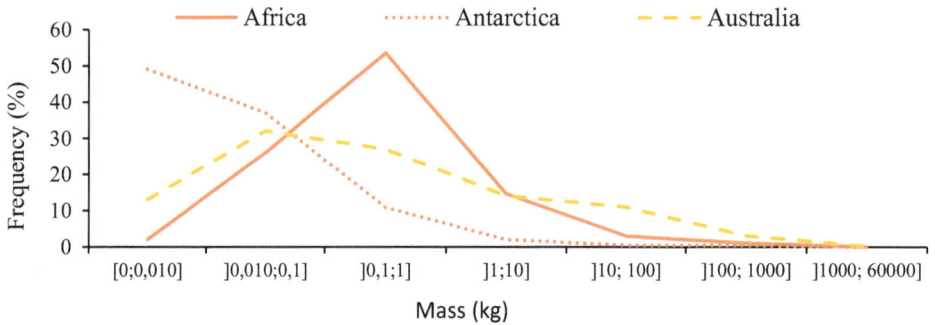

Fig. (10). Frequency distribution of meteorite finds masses in Africa, Australia and Antarcta. Data extracted from Meteoritical Bulletin Database.

In Africa, most of the lightweight meteorites are distributed in the northern hemisphere, while most of the heaviest ones are collected in the southern hemisphere. The map with geographic coordinates of the spatial distribution of 2751 meteorites, out of 12511, in Africa. Fig. (**11**) shows that iron meteorites account for almost 85% of the mass of all African finds so far (against 90% worldwide), with stony and stony-iron meteorites representing 15% and 0.5%, respectively.

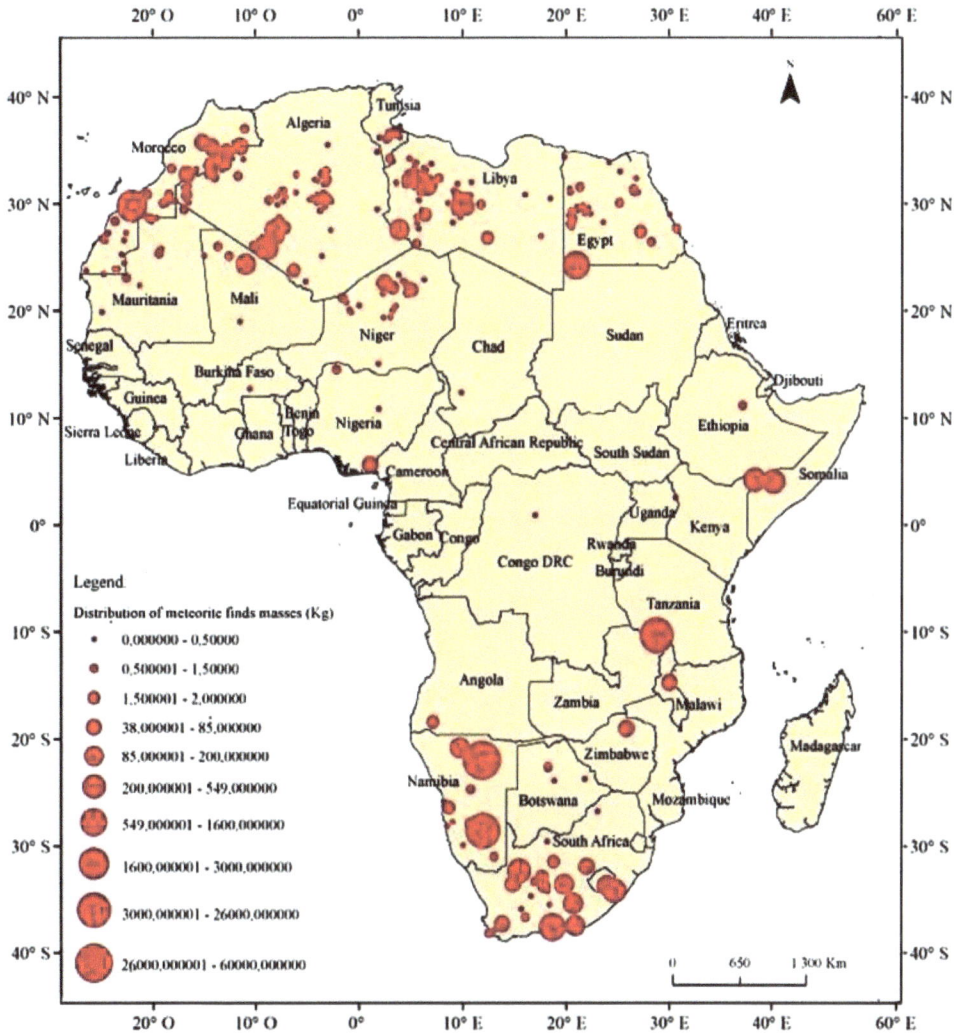

Fig. (11). Distribution map of meteorite finds in Africa as a function of mass.

The highest masses of iron meteorites can be ascribed to their higher resistance to atmospheric fragmentation with respect to chondrites and achondrites (Lee *et al.*, 2017). Furthermore, these large fragments are easier to be found, whereas the total mass of all fragments collected is supposed to be close to the mass of the body before its fragmentation (Hughes, 1992). Hawkins (1960) was the first to worry about the difficulty of collecting very small meteorites.

GRADES AND FACTORS OF WEATHERING OF AFRICAN METEORITE FINDS

Weathering Grades

From the moment a meteorite enters the earth atmosphere is exposed to a range of weathering agents, including water, air, salts, wind and temperature variations, so it becomes subject to several weathering processes in the terrestrial environment (Bland *et al.*, 2006). Other factors, such as the composition and porosity of the meteorite and host soil, may also influence the mechanisms and rates of weathering (Al Kathiri *et al.*, 2005). Various approaches and methods have been attempted to quantify meteorite alteration, including mineralogical analysis (Al-Kathiri *et al.*, 2005; Lee and Bland, 2004), quantification of chemical contamination (Al-Kathiri *et al.*, 2005; Hezel *et al.*, 2011), quantification of the oxidation degree by Mössbauer spectroscopy (Bland *et al.*, 1998) and iron isotope analyses (Saunier *et al.*, 2010). As a result of these studies, six weathering grades (from W1 to W6) were fixed (Wlotzka, 1993; Al-Kathiri *et al.*, 2005). In general, meteorite weathering can be regarded as the alteration of original component phases to phases that are more stable on Earth's surface. The alteration often starts with minor to complete oxidation of metals and then affects troilite (categories W1–W4) and continues with minor (W5) and then massive (W6) alteration of mafic silicates.

More than 99.5% of the 12511 meteorite finds recovered in Africa between 1710 and 2020, *i.e.*, 12448 samples, have been collected in the North, and 0.65% (64 finds) in the tropical zone and the South of the continent. Data collected on 9797 samples show that all weathering grades from W0 to W6 are covered. In particular, the distribution of weathering grades Fig. (**12**) shows the dominance of level W2 (44%), which proves a good preservation, whereas 5% of the collection features a weathering grade of W4, and only a few samples show W5 and W6 grades. Some differences exist in the alteration between samples collected in the North and those collected in the tropical zone and South of the continent. The presence of different weathering grades (W0 to W6) in the North indicates that in these areas, meteorites survive for long periods, which explains the presence of all weathering rates. The fact that meteorites recovered in the tropical zone and southwest of the continent, mostly ordinary chondrites, present weathering grades up to W4 is probably due to the disappearance of some of them because they are subject to complete weathering, whereas the remained ones probably have recent terrestrial age.

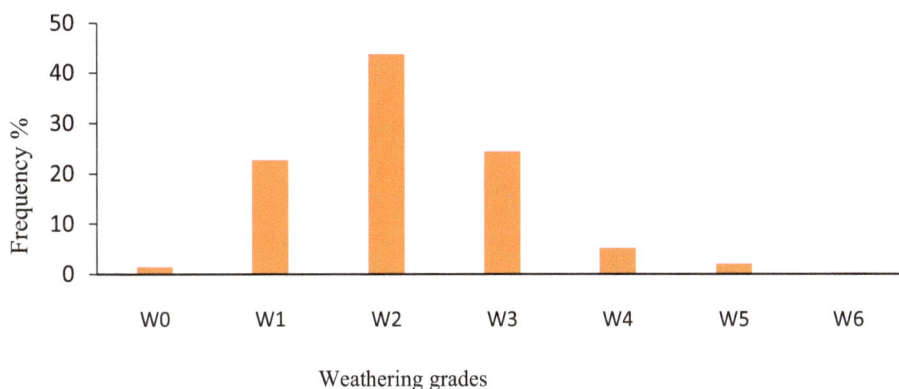

Fig. (12). Weathering grade frequency of meteorite finds in Africa.

The frequency of weathering of African, Australian, and Antarctic meteorite populations Fig. (**13**) shows that the distribution peak, which should represent the weathering grade required before a meteorite would begin to be seriously altered and then lost, in principle, should be the same for all three populations. However, the African meteorite finds feature less pronounced weathering, with a peak at W2, than the Australian collection, with a peak at W1, but more weathering than the Antarctic finds that shows a peak at W3. These results indicate that severe meteorite alteration has not yet started for the African population (68% of samples have a weathering grade under W3).

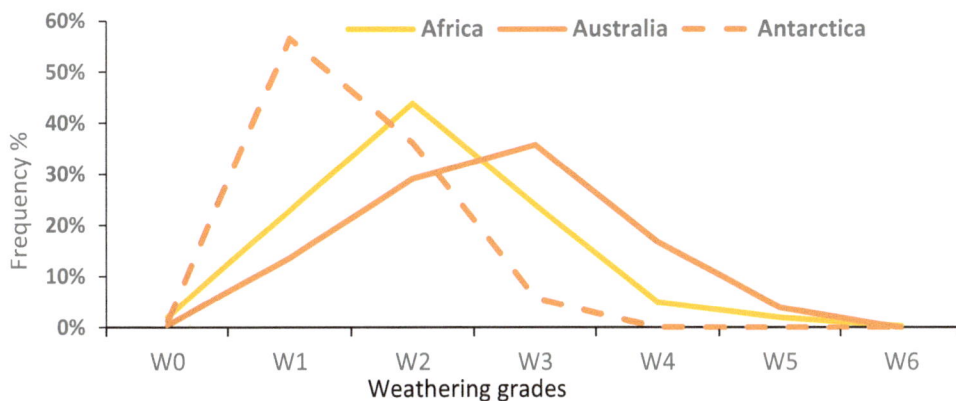

Fig. (13). Weathering grades of African, Australian and Antarctic meteorites.

According to Koeberl and Cassidy (1991), the Antarctic population shows less weathering because it has remained buried in ice and is subjected to an extremely conserving climate. The alteration of African finds may be related to the intrinsic characteristics (porosity, for example) and/or terrestrial ages of the dominant type samples, and also to the climates (dry or humid) of their recovery locations.

Factors Influencing the Weathering Grade

Climate

Over the entire earth's surface, meteorites have been exposed to a wide range of climatic changes. Searches conducted in arid areas, like the African deserts (Bischoff and Greiger, 1995; Schlüter *et al.*, 2002; Ibhi, 2013; Ibhi, 2014), Nullarbor in Australia (Bland *et al.*, 1996) and Antarctica (Righter *et al.*, 2014) have been very successful in meteorite finds. This is because in dry climates, chemical weathering is less vigorous, and this process might be further reduced where meteorites accumulate (Bland *et al.*, 1996).

The distribution of meteorite finds in Africa as a function of type is shown in Fig. (**14**). The database adopted includes 2751 meteorites (out of 12511) with geographic coordinates. Using the ArcGIS 10.2.2 application, the projection of these coordinates shows that most productive meteorite finds locations in Africa are concentrated in regions with a desert climate, *i.e.*, in the North, mainly in semi-arid to arid regions of Morocco to Egypt, and in the northern border of the Sahara Desert (Mauritania, Mali and Niger). Thus, despite the multiple wetter periods that occurred during the past 20000 years, meteorites were well preserved in the hot deserts of Africa (Bland *et al.*, 1998). In the South of the equator, the finds were collected in southwest areas, especially in the Namib and Kalahari deserts, which are characterized by a semi-arid to arid climate with precipitations too low to produce any vegetation (Maldague, 2006), which allowed the preservation of meteorites over a long time. For example, the majority (84%) of samples recovered in Libya's Dar al Gani region, the densest meteorite collection area in Africa, feature weathering grades lower than W3. The typical yearly rainfall in this region is between 10 and 20 mm, and an underground karst system rapidly removes rainwater (Schlüter *et al.*, 2002), which is also lost by intense evapotranspiration (Laity, 2009). Furthermore, the Dar al Gani desert pavement consists of limestone so that ablation by wind-blown limestone dust is significantly less abrasive than quartz-sand dust (Schlüter *et al.*, 2002).

Differently, in the tropical regions where two types of climate dominate, *i.e.*, the one with constant equatorial humidity at low latitudes near to the equator, and the other with alternate dry and wet seasons, which includes the regions affected by Asian monsoons or by high-pressure centers (Maldague, 2006), the number of meteorite finds is very low (15 samples). In these regions, meteorites weathering is caused by intense precipitation rates with averages exceeding 1000 mm, *e.g.*,

1830 mm/year in Gabon, 1646 mm/year in Congo, 1535 mm/year in Cameroon and 2177 mm/year in Equatorial Guinea, and high humidity, which determines highly destructive conditions that rapidly disaggregate meteorites, mainly those of stony type. The distribution Fig. (**14**) shows that iron meteorites are more resistant to climate effects in the southern hemisphere of Africa. The rare samples of stony meteorites recovered are ordinary chondrites without the absence of achondrites that are vulnerable to alteration by the Indian humid winds.

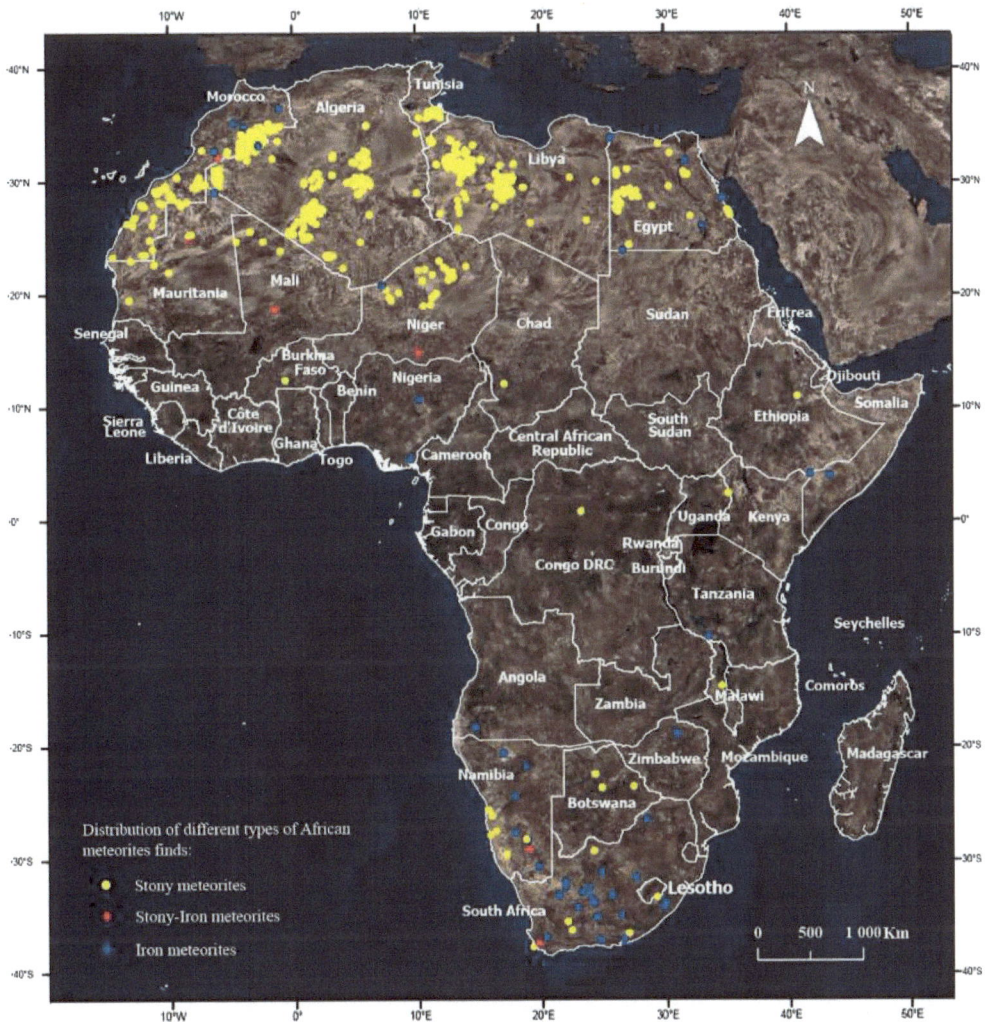

Fig. (14). Distribution map of different types of meteorites in Africa.

These data show that climate impacts the distribution of meteorite finds in Africa as in other regions such as Australia (Bland, 1998; Lee & Bland, 2004) and Chile (Gattacceca *et al.*, 2011). In particular, liquid H_2O and Cl are considered the most active factors in the weathering process of meteorites (Buchwald, 1975). Meteorites from these regions appear to have continued to absorb water after their fall. For example, a few weeks after the Orgueil meteorite fall, Cloz (1864) found that it contained 9.06 wt% water, and nearly a century later, 19.89 wt% water was measured (Wiik, 1956). Compared to other meteorite types, ordinary chondrites are more prone to terrestrial alteration (Bland *et al.*, 2006), which explains their very modest abundance in African tropical zones and in the South of the continent. For example, Stelzner *et al.* (1999) found an increasing water content with increasing age and weathering grade in 3 H5 and 3 L6 chondrites from Reg Al Açfer (Algeria). Furthermore, the absence of carbonaceous chondrites may be due to their fast weathering as they contain a large amount of fine-grained matrix enriched in volatile elements and abundant organic compounds (Bland *et al.*, 2006). Finally, the iron meteorite/stone meteorite ratio of African finds is much higher (0.97) in tropical areas and the South, in comparison to the North of the continent (0.01). Thus, metallic meteorites are more resistant and generally subsist longer in these environments, where they are also significantly easier to be identified.

The alteration of meteorites also depends on the climate history since the fall occurred. For example, African north hot deserts have a complex paleoclimate history as they have been subjected to multiple cycles of drought and wet periods over the past 40000 years (Bland *et al.*, 1998). For example, Schlüter *et al.* (2002) suggested that the meteorites fallen in Dar al Gani during earlier arid periods may have been covered by soil, and so preserved during wet periods. In conclusion, the geographic distribution of African meteorite finds suggests that alteration is mostly caused by climate. In the persistent dry climates of North of Africa, meteorites are conserved and feature a low weathering grade, whereas in temperate and tropical African climates, meteorites are strongly affected by weathering. However, the few samples of recent fall that were collected in the South have a low weathering grade.

Sample Mass

Differently from Antarctic and Australian populations, the majority of African meteorite finds are medium to large samples with masses between 10 g and 1 kg (80%) and between 100 g and 1 kg (54%), which helps to conserve the collection. In general, the mass of finds depends on the size of fallen meteorites and of the pieces resulting from their impact (Hughes, 1992), as well as on the collected masses biased by the poor recovery efficiency of the very small ones (Korotev,

2012). In particular, stony meteorites are the most frequent finds (97.41%), with 80.29% chondrites, which indicates that the population of African finds is quite homogeneous and dominated by medium-sized stony meteorites.

The percentage of African finds featuring a weathering grade higher than W2 plotted as a function of the sample mass Fig. (**15**) shows that the percentage decreases with the mass increase, which proves that the larger is the mass, the less weathered the meteorite is. Hutzler *et al.* (2016) found a similar result for the Atacama collection in Chile, and suggested that a lower surface-to-volume ratio implies a lower weathering of a meteorite. Furthermore, Bischoff and Geiger (1995) suggested that "the degree of weathering varies on a cm-scale, thus the effect of weathering decreases from the surface to the center of the meteorite". This implies that the outer mms of a meteorite can be severely weathered (W4), while at 1-cm depth, the degree of weathering can be moderate (W2). According to Zolensky *et al.* (2006), the major effects of weathering are an increase in the number of samples, a decrease in the population of all meteorites, and a reduction in the mass of survivors. Thus, the removal of small-sized meteorites from a population at various rates of alteration can change the mass distribution.

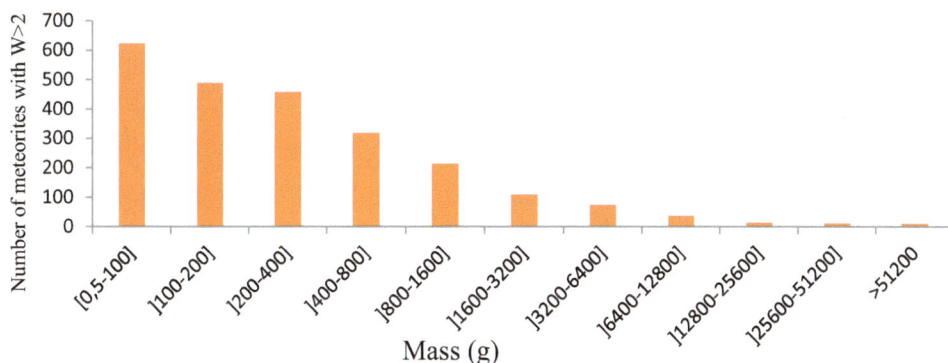

Fig. (15). The number of meteorite finds in Africa featuring a weathering grade greater than W2 as a function of the sample mass.

Terrestrial Age

The terrestrial age, *i.e.*, the time the meteorite has been on Earth since it fell, is an essential factor in identifying a meteorite's history (Jull, 2001). The average terrestrial residence time of Antarctic meteorites is substantially longer compared to non-Antarctic meteorites (Grady *et al.*, 1991), which, particularly stony meteorites, are destroyed by erosion and weathering more quickly than in Antarctica (Bland *et al.*, 2012).

According to Bland *et al.* (1998), the terrestrial residence of meteorites in the North of Africa is less than 20000 years. Indeed, the distribution of the terrestrial age of 91 finds from Africa for which the residence age is known Fig. (**16**) shows that most of them (80%) have a terrestrial age below 20000 years, with 50% of total samples featuring an age less than 10000 years, while the remaining samples feature an age ranging between 20000 and 85000 years. Actually, meteorites found in northern hot deserts have younger terrestrial ages, with 92% of them being younger than 20000 years in the Daraje DCA in Libya (Jull *et al.*, 1990), 83% in the Reg el Açfer DCA in Algeria (Wlotzka, 1995), and 70% in the Dar El Gani DCA in Libya (Welten *et al.*, 2004). The African distribution is significantly similar to that of Australian meteorites, 44-85% of which have a terrestrial age lower than 10000 years and only 0-18.5% more than 30000 years (Bland *et al.*, 1998). However, the distribution ages of African finds are different from that of other areas, *e.g.*, the Oman population includes only 16% of meteorites with terrestrial ages lower than 10000 years and 28% with ages superior to 30000 years (Al Kathiri *et al.*, 2005). Only 42 of the 91 meteorites in Fig. (**16**) feature a weathering grade ranging between W1 and W5 with a prevalence (60%) of W2 and W3 (Fig. **17**).

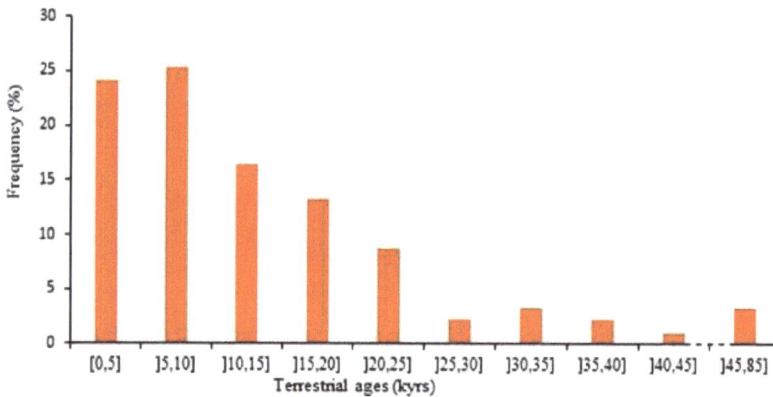

Fig. (16). Terrestrial age frequency of 91 meteorite finds in Africa. Data Source: Jull *et al.*, 1990; Welten *et al.*, 2004; and Jull, 2010.

Several studies (Al-Kathiri *et al.* 2005; Zurfluh *et al.* 2011; Korotev, 2012) have indicated that the degree of chemical weathering of meteorites rises with increasing their terrestrial age on the dependence on their concentrations of Sr. The comparison of weathering grade with terrestrial age of meteorites recovered from the hot desert in Roosevelt County in the western USA yielded the following results: W2: 500-15000 years, W3: 15000-30000 years, W4: 20000-35000 years, W5 and W6: 30000 to > 45000 years (Norton, 2002). African meteorite population shows a similar rise of weathering grades versus terrestrial age (Fig.

17), but no clear correlation could be established because of the limited number of data available. However, meteorites with a weathering grade W2 feature a terrestrial age of 2500-16300 years, those of grade W3 an age from 3500 to 29100 years, and those of grade W4 range from 5000 to 15100 years. These values may be ascribed to the humid environment typical of north Africa in the period ranging from 16000 to 13000 years before the semiarid early Holocene characterized by rainfall peaks between about 6000-5000 and 2000 years (Shaw *et al.*, 1986; Jull, 2001). In conclusion, African meteorite finds can be regarded as a young and slightly altered population that exhibits an apparent correlation between the weathering grade and terrestrial age based on the small proportion (32%) of samples with a weathering grade higher than W2.

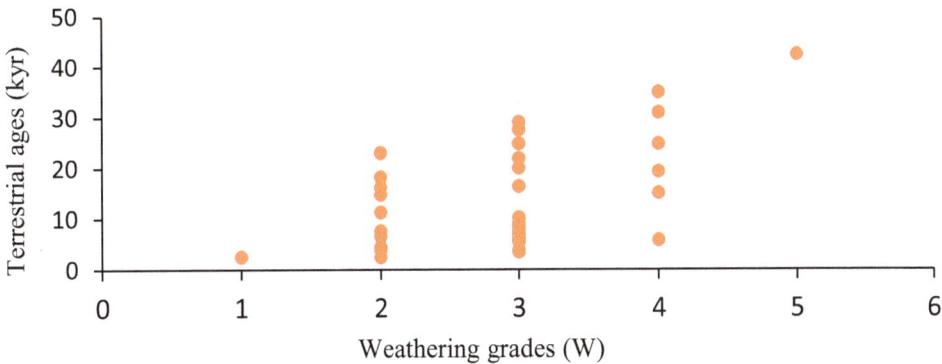

Fig. (17). Weathering grades meteorite finds in Africa as a function of their terrestrial age. Data Source: Jull *et al.*, (1990), Welten *et al.*, (2004), and Jull (2010).

Sample Porosity

The degree of porosity or fracturing is one of the main factors controlling the processes and grades of weathering of meteorites (Velbel *et al.*, 1991). According to Coulson *et al.* (2007), porosity is influenced by shock effects associated with the collision and/or ejection processes. Furthermore, the shock stage determines the degree of fracturing of the chondrite meteorite matrix (Stöffler *et al.*, 1991). The rock is heated, melted and deformed (shock metamorphism) by the pressure caused by the meteoroid's impact on its parent body. To evaluate the level of shock metamorphism, meteorites are typically classified in a shock stage classification between S1 and S6: unshocked (S1); very weakly shocked (S2); weakly shocked (S3); moderately shocked (S4); strongly shocked (S5); and very strongly shocked (S6) (Sharp & DeCarli, 2006). This classification depends mainly on the origin of pores and their history of shock (Pesonen *et al.*, 1993). According to several studies, the porosity of ordinary chondrites is less than 10% (Consolmagno *et al.*, 1998; El Abbasi *et al.*, 2013), and is mainly controlled by shock, with a sharp decrease of porosity between shock stages S2 and S3 (Gattacceca *et al.*, 2005). D'Orazio *et al.* (2011) reported that a system of progressive shock metamorphisms occurs in impactites consisting of silicate rocks and sediments. The "shock-metamorphosed rocks" appear as lithic clasts or melt particles in impactites close to craters and infrequently in distant impactites, which feature an extensive variety of metamorphism ranging from unshocked to shock melted. However, as the degree of shock metamorphism at a given shock pressure mostly depends on the mineralogical composition and porosity of the rock or sediment, distinct shock categorization methods are needed for various types of planetary rocks and sediments (Stöffler *et al.*, 2018).

To evaluate the relation between the porosity of African finds and their weathering grade, the shock stage is considered a proxy for the original porosity of meteorites (Ouknine *et al.*, 2019). African meteorites include 49% of samples with shock stages ranging from S0 to S2 and 51% from S3 to S6 in. The frequency of weathering grade as a function of the shock stage of African meteorites Fig. (**18**) shows that 52% of meteorites with weathering grade W2 have shock stages between S0 and S2 and 48% have shock stages from S3 to S6, whereas 60% of meteorites with low weathering grades, *i.e.*, W0 and W1, have shock stages ranging from S3 to S6. Noteworthy, the percentage of samples with shock stages S0-S2 rises with the increase of weathering grades from W3 to W5, and all meteorites with W6 have a shock stage S0-S2. Therefore, meteorites with higher initial porosity, *i.e.*, about 10%, are mostly more weathered than meteorites with lower initial porosity, *i.e.*, about 5%. According to Consolmagno *et al.* (1998, 2008), more porous meteorites are vulnerable to faster weathering because water can enter the sample more easily. Thus, a major variable that affects the alteration of a meteorite population is its initial porosity.

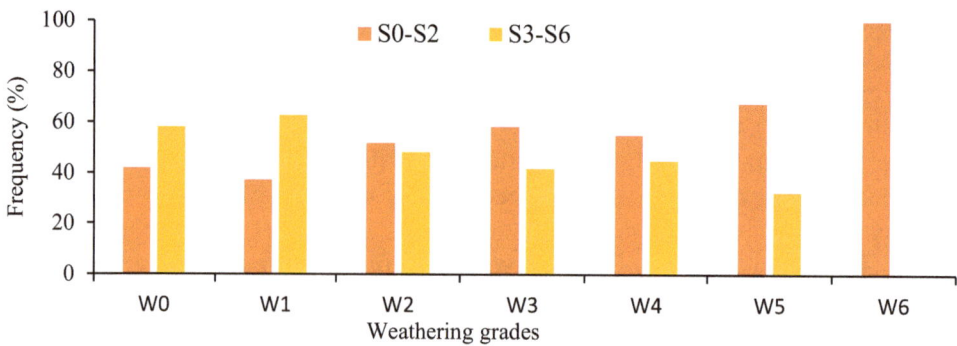

Fig. (18). Frequency distribution of weathering grades as a function of shock stages for African meteorite finds.

SUMMARY AND CONCLUSION

The high rate of meteorite discoveries in Africa does not appear to be related to the total population density and its distribution, but depends on the group of people interested in these extraterrestrial rocks, such as nomads, meteorite hunters and scientists, who have the ability to distinguish meteorites from terrestrial rocks. In Africa, the majority of meteorite research occurs in dry, hot deserts with very little vegetation during the humid months of autumn and spring, especially in April.

The classification of African meteorite finds indicates that this population includes all known meteorite types, with a significant proportion of rare and abundant stony meteorites. In the last three decades, a decrease in the frequencies of chondrite finds, particularly ordinary chondrites, and a significant rise in carbonaceous chondrites and achondrites have been detected. This may be due to the rise in the value of carbonaceous and achondritic meteorites, in comparison to chondritic meteorites, in the meteorite markets in North Africa. The estimated decrease of stony-iron and iron meteorite finds could be due to the wide recoveries that occurred in prehistorical and historical times.

The African population includes meteorite finds featuring a mass peak of 1 kg, with the majority of lightweight meteorites discovered in the northern hemisphere of the continent and the heaviest meteorites found in the southern hemisphere. The African collection shows a mass distribution typical of areas searched on foot by nomads or by car, mostly by private collectors and mineral dealers in desert areas of Morocco, Algeria and Libya.

The African meteorite finds feature weathering grades ranging from W0 to W6, with 68% of samples showing a weathering grade below W3, which proves a good preservation of meteorites. The factors influencing the mechanisms and rates of weathering of African meteorite population include: (i) *climate*, which is the main factor allowing the preservation and accumulation of meteorites in North and southwest Africa (Namib and Kalahari deserts) due to the semi- arid, arid or hyperarid conditions, whereas, meteorites disappear in the tropical and southeast regions of the continent because of the wet and rainy conditions; (ii) *sample mass*, the variation of which features a clear relationship with the meteorite grade of alteration (superior to W2), proving that small samples are preferentially more altered than large ones; (iii) *sample terrestrial age*, which appears to be correlated with the degree of weathering, showing that samples collected in this continent are less weathered due to their relatively young age, with 80% of them having a terrestrial age of less than 20000 years; and (iv) *sample porosity*, showing that meteorites with higher initial porosity (S0-S2), *i.e.*, 49% of African finds, are more weathered than meteorites with lower initial porosity (from S3 to S6), *i.e.*, 51%. In conclusion, the prevalence of medium-mass samples (1 kg) in the African collection dominated by stony meteorites can be ascribed to the younger age, low alteration with a prevalent weathering grade W2, and low initial porosity. These characteristics are clearly related to the arid climate predominant in Africa. Actually, the majority of recoveries (99%) occur in north hot deserts, which facilitates meteorite preservation and accumulation, and also allows an easier discovery. Due to this, the African continent is one of the most prolific areas for meteorites recovery, accounting for more than 1/6 of all recoveries worldwide.

REFERENCES

Agee, C., Bullock, E., Bouvier, A., Dunn, T., Gattacceca, J., Grossman, J., Herd, C., Ireland, T., Metzler, K., Mikouchi, T., Ruzicka, A., Smith, C., Welten, K., Welzenbach, L. (2015). Categorization of Finds and Falls. Available at: http://www.lpi.usra.edu/meteor/docs/falls-finds.pdf(Accessed on: 2017).

Al-KATHIRI, A., Hofmann, B.A., Jull, A.J.T., Gnos, E. (2005). Weathering of meteorites from Oman: Correlation of chemical and mineralogical weathering proxies with ^{14} C terrestrial ages and the influence of soil chemistry. *Meteorit. Planet. Sci., 40*(8), 1215-1239. [http://dx.doi.org/10.1111/j.1945-5100.2005.tb00185.x]

Bevan, A.W.R., Bland, P.A., Jull, A.L.T. (1998). Meteorite flux on the Nullarbor Region, Australia. *Meteorites: Flux with Time and Impact Effects. Special Publications 140.* (pp. 59-73). London, U.K: Geological Society.

Bevan, A.W.R. (1992). Australian meteorites *Records of the Australian Museum Supplement., 15*(1), 1-27. [http://dx.doi.org/10.3853/j.0812-7387.15.1992.80]

Bischoff, A., Geiger, T. (1995). Meteorites from the Sahara: Find locations, shock classification, degree of weathering and pairing. *Meteorit. Planet. Sci., 30*(1), 113-122.

Bischoff, A., Weber, D. (1997). Dar al Gani 262: the first lunar meteorite from the Sahara. *Meteorit. Planet. Sci., 32*(4), A13.

Bland, P.A., Berry, F.J., Smith, T.B., Skinner, S.J., Pillinger, C.T. (1996). The flux of meteorites to the Earth and weathering in hot desert ordinary chondrite finds. *Geochim. Cosmochim. Acta, 60*(11), 2053-2059. [http://dx.doi.org/10.1016/0016-7037(96)00125-1]

Bland, P.A., Bevan, A.W.R., Jull, A.J.T. (2000). Ancient meteorite finds and the Earth's surface environment. *Quat. Res., 53*(2), 131-142. [http://dx.doi.org/10.1006/qres.1999.2106]

Bland, P.A., Sexton, A.S., Jull, A.J.T., Bevan, A.W.R., Berry, F.J., Thornley, D.M., Astin, T.R., Britt, D.T., Pillinger, C.T. (1998). Climate and rock weathering: a study of terrestrial age dated ordinary chondritic meteorites from hot desert regions. *Geochim. Cosmochim. Acta, 62*(18), 3169-3184. [http://dx.doi.org/10.1016/S0016-7037(98)00199-9]

Bland, P.A., Spurný, P., Bevan, A.W.R., Howard, K.T., Towner, M.C., Benedix, G.K., Greenwood, R.C., Shrbený, L., Franchi, I.A., Deacon, G., Borovička, J., Ceplecha, Z., Vaughan, D., Hough, R.M. (2012). The Australian Desert Fireball Network: a new era for planetary science. *Aust. J. Earth Sci., 59*(2), 177-187. [http://dx.doi.org/10.1080/08120099.2011.595428]

Bland, P. A., Zolensky, M. E., Benedix, G. K. & Sephton, M. A. (2006). Weathering of chondritic meteorites. Meteorites and the early solar system II, 853-867. [http://dx.doi.org/10.2307/j.ctv1v7zdmm.45]

Bland, P.A. (2001). Quantification of meteorite infall rates from accumulations in deserts, and meteorite accumulations on Mars. In: Peucker-Ehrenbrink, B., Schmitz, B., (Eds.), *Accretion of extraterrestrial matter*

throughout Earth's history. (pp. 267-303). Boston, MA: Springer.
[http://dx.doi.org/10.1007/978-1-4419-8694-8_15]

Bottke, W.F., Nesvorný, D., Grimm, R.E., Morbidelli, A., O'Brien, D.P. (2006). Iron meteorites as remnants of planetesimals formed in the terrestrial planet region. *Nature, 439*(7078), 821-824.
[http://dx.doi.org/10.1038/nature04536] [PMID: 16482151]

Buchwald, V. F. (1975). Handbook of iron meteorites: Their history, distribution, composition, and structure. Volumes 1, 2 & 3. Arizona: State University, Center for Meteorite Studies, and Berkeley: University of California Press.

Burbine, T.H., McCoy, T.J., Meibom, A., Gladman, B., Keil, K. (2002). Meteoritic parent bodies: Their number and identification. *Asteroids, III*, 653-668.
[http://dx.doi.org/10.2307/j.ctv1v7zdn4.50]

Cassidy, W.A., Rancitelli, L.A. (1982). Antarctic Meteorites: The abundant material being discovered in Antarctica may shed light on the evolution of meteorite parent bodies and the history of the solar system. *Am. Sci., 70*(2), 156-164.

Chladni, E. F. F. (1794). Ueber den Ursprung der von Pallas gefundenen und anderer ihr ähnlicher Eisenmassen und über einige damit in Verbindung stehende Naturerscheinungen. Bey Johann Friedrich Hartknoch.

Ciesla, F.J. (2005). In: Krot, A.N., Scott, E.R.D., Reipurth, B. (Eds.), Astronomical Society of the Pacific Conference Series: Chondrites and the Protoplanetary Disk. vol. 341.

Cloëz, S. (1864). Analyse chimique de la pierre météorique d'Orgueil. *Comptes Rendus de l'Académie des Sciences, 59*, 37-40.

Consolmagno, G.J., Britt, D.T., Macke, R.J. (2008). The significance of meteorite density and porosity. *Chem. Erde, 68*(1), 1-29.
[http://dx.doi.org/10.1016/j.chemer.2008.01.003]

Consolmagno, G.J., Britt, D.T., Stoll, C.P. (1998). The porosities of ordinary chondrites: Models and interpretation. *Meteorit. Planet. Sci., 33*(6), 1221-1229.
[http://dx.doi.org/10.1111/j.1945-5100.1998.tb01307.x]

Corrigan, C.M., Welzenbach, L.C., Righter, K., McBride, K.M., McCoy, T.J., Harvey, R.P., Satterwhite, C.E. (2014). Chapter 10: A statistical look at the U.S. Antarctic meteorite collection. In: Righter, K., Harvey, R. P., Corrigan, C. M., McCoy, T. J., (Eds.), *Thirty-five Seasons of U.S. Antarctic Meteorites.* Special Publications 68, American Geophysical Union.(Vol. 296, pp. 101-130). Washington, D. C.:

Corti, G., Zeoli, A., Bonini, M. (2003). Ice-flow dynamics and meteorite collection in Antarctica Earth and Planetary Science Letters. *215*, 371-378.

Coulson, I.M., Beech, M., Nie, W. (2007). Physical properties of Martian meteorites: Porosity and density measurements. *Meteorit. Planet. Sci., 42*(12), 2043-2054.
[http://dx.doi.org/10.1111/j.1945-5100.2007.tb01006.x]

DeMeo, F.E., Alexander, C.M.O.D., Walsh, K.J., Chapman, C.R., Binzel, R.P. (2015). The compositional structure of the asteroid belt. *Asteroids IV, 1*, 13.
[http://dx.doi.org/10.2458/azu_uapress_9780816532131-ch002]

Dennison, J.E., Lingner, D.W., Lipschutz, M.E. (1986). Antarctic and non-Antarctic meteorites form different populations. *Nature, 319*(6052), 390-393.
[http://dx.doi.org/10.1038/319390a0]

D'Orazio, M., Folco, L., Zeoli, A., Cordier, C. (2011). Gebel Kamil: The iron meteorite that formed the Kamil crater (Egypt). *Meteorit. Planet. Sci., 46*(8), 1179-1196.
[http://dx.doi.org/10.1111/j.1945-5100.2011.01222.x]

El Abassi, D., Ibhi, A., Faiz, B., Aboudaoud, I. (2013). A simple method for the determination of the porosity and tortuosity of meteorites with ultrasound. *Journal of Geophysics and Engineering, 10*(5), 055003.

[http://dx.doi.org/10.1088/1742-2132/10/5/055003]

FAO, (2002). Evaluation des Ressources Forestières Mondiales. Rome, 140.

Gattacceca, J., Rochette, P., Denise, M., Consolmagno, G., Folco, L. (2005). An impact origin for the foliation of chondrites. *Earth Planet. Sci. Lett., 234*(3-4), 351-368.
[http://dx.doi.org/10.1016/j.epsl.2005.03.002]

Gattacceca, J., Valenzuela, M., Uehara, M., Jull, A.J.T., Giscard, M., Rochette, P., Braucher, R., Suavet, C., Gounelle, M., Morata, D., Munayco, P., Bourot-Denise, M., Bourles, D., Demory, F. (2011). The densest meteorite collection area in hot deserts: The San Juan meteorite field (Atacama Desert, Chile). *Meteorit. Planet. Sci., 46*(9), 1276-1287.
[http://dx.doi.org/10.1111/j.1945-5100.2011.01229.x]

Gounelle, M., (2017). Les météorites, Que sais-je? 2ème édition n° 3859, Presses universitaires de France.

Grady, M.M., Wright, I.P., Pillinger, C.T. (1991). Comparisons between Antarctic and non-Antarctic meteorites based on carbon isotope geochemistry. *Geochim. Cosmochim. Acta, 55*(1), 49-58.
[http://dx.doi.org/10.1016/0016-7037(91)90398-O]

Hawkins, G.S. (1960). Asteroidal fragments. *Astron. J., 65*, 318.
[http://dx.doi.org/10.1086/108251]

Herbst, W., Greenwood, J.P. (2019). A radiative heating model for chondrule and chondrite formation. *Icarus, 329*, 166-181.
[http://dx.doi.org/10.1016/j.icarus.2019.03.039]

Hezel, D.C., Schlüter, J., Kallweit, H., Jull, A.J.T., Al Fakeer, O.Y., Al Shamsi, M., Strekopytov, S. (2011). Meteorites from the United Arab Emirates: Description, weathering, and terrestrial ages. *Meteorit. Planet. Sci., 46*(2), 327-336.
[http://dx.doi.org/10.1111/j.1945-5100.2010.01165.x]

Hughes, D. (1992). The meteorite flux. *Space Sci. Rev., 61*(3-4), 275-299.
[http://dx.doi.org/10.1007/BF00222309]

Huss, G.R. (1991). Meteorite mass distributions and differences between Antarctic and non-Antarctic meteorites. *Geochim. Cosmochim. Acta, 55*(1), 105-111.
[http://dx.doi.org/10.1016/0016-7037(91)90404-S]

Hützler, A., Gattacceca, J., Rochette, P., Braucher, R., Carro, B., Christensen, E.J., Cournede, C., Gounelle, M., Laridhi Ouazaa, N., Martinez, R., Valenzuela, M., Warner, M., Bourles, D. (2016). Description of a very dense meteorite collection area in western Atacama: Insight into the long-term composition of the meteorite flux to Earth. *Meteorit. Planet. Sci., 51*(3), 468-482.
[http://dx.doi.org/10.1111/maps.12607]

Ibhi, A. (2013). New Mars meteorite fall in Morocco: final strewn field. International Letters of Chemistry. *Physics and Astronomy, 11*, 20-25.

Ibhi, A. (2014). Tighert: a new eucrite meteorite fall from morocco. *International Meteor Conference, 33*, 1-3.

Ibhi, A. (2016). *Météorites : Perles du Désert Marocain.*

Johnson, B.C., Minton, D.A., Melosh, H.J., Zuber, M.T. (2015). Impact jetting as the origin of chondrules. *Nature, 517*(7534), 339-341.
[http://dx.doi.org/10.1038/nature14105] [PMID: 25592538]

Jull, A.J.T., Wlotzka, F., Palme, H., Donahue, D.J. (1990). Distribution of terrestrial age and petrologic type of meteorites from western Libya. *Geochim. Cosmochim. Acta, 54*(10), 2895-2898.
[http://dx.doi.org/10.1016/0016-7037(90)90028-J]

Jull, A.J.T., McHARGUE, L.R., Bland, P.A., Greenwood, R.C., Bevan, A.W.R., Kim, K.J., LaMOTTA, S.E., Johnson, J.A. (2010). Terrestrial ages of meteorites from the Nullarbor region, Australia, based on 14C and 14C-10Be measurements. *Meteorit. Planet. Sci., 45*(8), 1271-1283.

[http://dx.doi.org/10.1111/j.1945-5100.2010.01289.x]

Jull, A.T. (2001). Terrestrial ages of meteorites. *Accretion of extraterrestrial matter throughout Earth's history.* Springer.(pp. 241-266). Boston, MA:
[http://dx.doi.org/10.1007/978-1-4419-8694-8_14]

Khiri, F., Ibhi, A., Saint-Gerant, T., Medjkane, M., Ouknine, L. (2017). Meteorite falls in Africa. *J. Afr. Earth Sci., 134,* 644-657.
[http://dx.doi.org/10.1016/j.jafrearsci.2017.07.022]

Koeberl, C., Cassidy, W.A. (1991). Differences between Antarctic and non-Antarctic meteorites: An assessment. *Geochim. Cosmochim. Acta, 55*(1), 3-18.
[http://dx.doi.org/10.1016/0016-7037(91)90395-L]

Korotev, R.L. (2012). Lunar meteorites from Oman. *Meteorit. Planet. Sci., 47.*

Korotev, R.L. (2017). Some Meteorite Statistics. Available at: https://sites.wustl.edu/meteoritesite/items/some-meteorite-statistics/(Accessed on: 2019).

Krot, A.N., Amelin, Y., Cassen, P., Meibom, A. (2005). Young chondrules in CB chondrites from a giant impact in the early Solar System. *Nature, 436*(7053), 989-992.
[http://dx.doi.org/10.1038/nature03830] [PMID: 16107841]

Krot, A.N., Keil, K., Scott, E.R.D., Goodrich, C.A., Weisberg, M.K. (2014). Classification of meteorites and their genetic relationships. *Meteorites, Comets, and Planets, 1,* 1-63.
[http://dx.doi.org/10.1016/B978-0-08-095975-7.00102-9]

Laity, J. J. (2009). Deserts and desert environments (Vol. 3). John Wiley & Sons, Inc., Publication.

Larouci, N., Aoudjehane, H.C., Jambon, A. (2014). Méthodologie d'étude des météorites du Maroc Methodology of studying meteorites from Morocco. *Bulletin de l'Institut Scientifique, Rabat,* (36), 69-83.

Lee, M.R., Bland, P.A. (2004). Mechanisms of weathering of meteorites recovered from hot and cold deserts and the formation of phyllosilicates. *Geochim. Cosmochim. Acta, 68*(4), 893-916.
[http://dx.doi.org/10.1016/S0016-7037(03)00486-1]

Lee, N., Fritz, J., Fries, M.D., Gil, J.F., Beck, A., Pellinen-Wannberg, A., Schmitz, B., Steele, A., Hoffmann, B.A. (2017). The extreme biology of meteorites: Their role in understanding the origin and distribution of life on earth and in the universe. In: Stan-Lotter, H., Fendrihan, S., (Eds.), *Adaption of microbial life to environmental extremes.* Springer International Publishing.
[http://dx.doi.org/10.1007/978-3-319-48327-6_11]

MacPherson, G.J. (2014). Calcium-aluminum-rich inclusions in chondritic meteorites. *Meteorites and cosmochemical processes,* 139-179.
[http://dx.doi.org/10.1016/B978-0-08-095975-7.00105-4]

Maldague, M. (2006). Traité de gestion de l'environnement tropical. Bibliothèque Paul-Émile Boulet de l'Université du Québec à Chicoutimi. 350.

Marchand, J. (1996). L'économie minière en Afrique australe, Karthala Editions, Paris, 424.

Meteoritical bulletin database, (2017). The meteoritical society, Available at: http://www.lpi.usra.edu/ Accessed on January, 2017.

Morbidelli, A., Lunine, J.I., O'Brien, D.P., Raymond, S.N., Walsh, K.J. (2012). Building terrestrial planets. *Annu. Rev. Earth Planet. Sci., 40*(1), 251-275.
[http://dx.doi.org/10.1146/annurev-earth-042711-105319]

Norton, O. R. (2002). The Cambridge encyclopedia of meteorites, Cambridge University Press, Cambridge, 374.

Notkin, G. (2011). Meteorite Hunting: How to find treasure from space. Aerolite Meteorites, Stanegate Press, 318.

Pesonen, L.J., Terho, M., Kukkonen, I.T. (1993). Physical properties of 368 meteorites: implications from meteorite magnetism and planetary geophysics. *Proceeding NIPR Symposium Antarctica Meteorites 6,* 401-416.

Righter, K., Corrigan, C., McCoy, T. & Harvey, R. (2014). 35 Seasons of US Antarctic meteorites (1976-2010): A Pictorial Guide to the Collection, Wiley, J. & Sons Inc., Hoboken, New Jersey, 189.

Sanders, I.S., Scott, E.R.D. (2012). The origin of chondrules and chondrites: Debris from low-velocity impacts between molten planetesimals? *Meteorit. Planet. Sci., 47*(12), 2170-2192.
[http://dx.doi.org/10.1111/maps.12002]

Saunier, G., Poitrasson, F., Moine, B., Gregoire, M., Seddiki, A. (2010). Effect of hot desert weathering on the bulk-rock iron isotope composition of L6 and H5 ordinary chondrites. *Meteorit. Planet. Sci., 45*(2), 195-209.
[http://dx.doi.org/10.1111/j.1945-5100.2010.01017.x]

Schlüter, J., Schultz, L., Thiedig, F., Al-Mahdi, B.O., Aghreb, A.E.A. (2002). The Dar al Gani meteorite field (Libyan Sahara): Geological setting, pairing of meteorites, and recovery density. *Meteorit. Planet. Sci., 37*(8), 1079-1093.
[http://dx.doi.org/10.1111/j.1945-5100.2002.tb00879.x]

Scott, E.R.D. (1977). Pallasites—metal composition, classification and relationships with iron meteorites. *Geochim. Cosmochim. Acta, 41*(3), 349-360.
[http://dx.doi.org/10.1016/0016-7037(77)90262-9]

Sharp, T. G. & DeCarli, P. S. (2006). Shock effects in meteorites. Meteorites and the early solar system II 804.
[http://dx.doi.org/10.2307/j.ctv1v7zdmm.37]

Shaw, P.A., Cooke, H.J. (1986). Geomorphic evidence for the late Quaternary palaeoclimates of the middle Kalahari of northern Botswana. *Catena, 13*(4), 349-359.
[http://dx.doi.org/10.1016/0341-8162(86)90009-3]

Sorby, H.C. (1877). On the structure and origin of meteorites. *Nature, 15*(388), 495-498.
[http://dx.doi.org/10.1038/015495a0]

Stelzner, T.H., Heide, K., Bischoff, A., Weber, D., Scherer, P., Schultz, L., Happel, M., Schrön, W., Neupert, U., Michel, R., Clayton, R.N., Mayeda, T.K., Bonani, G., Haidas, I., Ivy-Ochs, S., Suter, M. (1999). An interdisciplinary study of weathering effects in ordinary chondrites from the Acfer region, Algeria. *Meteorit. Planet. Sci., 34*(5), 787-794.
[http://dx.doi.org/10.1111/j.1945-5100.1999.tb01391.x]

Stöffler, D., Hamann, C., Metzler, K. (2018). Shock metamorphism of planetary silicate rocks and sediments: Proposal for an updated classification system. *Meteorit. Planet. Sci., 53*(1), 5-49.
[http://dx.doi.org/10.1111/maps.12912]

Stöffler, D., Keil, K., Scott, E.R.D. (1991). Shock metamorphism of ordinary chondrites. *Geochim. Cosmochim. Acta, 55,* 3845-3867.
[http://dx.doi.org/10.1016/0016-7037(91)90078-J]

Velbel, M.A., Long, D.T., Gooding, J.L. (1991). Terrestrial weathering of Antarctic stone meteorites: Formation of Mg-carbonates on ordinary chondrites. *Geochim. Cosmochim. Acta, 55*(1), 67-76.
[http://dx.doi.org/10.1016/0016-7037(91)90400-Y]

Weisberg, M.K. (2017). Encyclopedia of Earth Sciences Series. In: White, W.M., (Ed.), *Encyclopedia of Geochemistry_ A Comprehensive Reference Source on the Chemistry of the Earth-Springer International Publishing.* 1-9.

Weisberg, M.K., McCoy, T.J., Krot, A.N. (2006). Systematics and evolution of meteorite classification. In: Lauretta, D.S., McSween, H.Y., (Eds.), *Meteorites and the Early Solar System II..* Tucson: University of Arizona Press. 19-52.
[http://dx.doi.org/10.2307/j.ctv1v7zdmm.8]

Welten, K.C., Nishiizumi, K., Finkel, R.C., Hillegonds, D.J., Jull, A.J.T., Franke, L., Schultz, L. (2004).

Exposure history and terrestrial ages of ordinary chondrites from the Dar al Gani region, Libya. *Meteorit. Planet. Sci., 39*(3), 481-498.
[http://dx.doi.org/10.1111/j.1945-5100.2004.tb00106.x]

Whillans, I.M., Cassidy, W.A. (1983). Catch a falling star: meteorites and old ice. *Science, 222*(4619), 55-57.
[http://dx.doi.org/10.1126/science.222.4619.55] [PMID: 17810091]

Wiik, H.B. (1956). The chemical composition of some stony meteorites. *Geochim. Cosmochim. Acta, 9*(5-6), 279-289.
[http://dx.doi.org/10.1016/0016-7037(56)90028-X]

Wlotzka, F. (1993). A weathering scale for the ordinary chondrites. *Meteoritics, 28*.

Wlotzka, F., Jull, A. J. T., Donahue, D. J. (1995). Carbone-14 terrestrial ages of meteorites from Açfer, Algeria. *LPI Technical Report., 95*(2), 72-73.

Zolensky, M., Bland, P., Brown, P., Halliday, I. (2006). Flux of extraterrestrial materials. *Meteorites and the early solar system II., 869-888*.
[http://dx.doi.org/10.2307/j.ctv1v7zdmm.46]

Zurfluh, F.J., Hofmann, B.A., Gnos, E., Eggenberger, U., Villa, I.M., Greber, N.D., Jull, A.J.T. (2011). New insights into the strontium contamination of meteorites outlook (abstract #5229). *Meteorit. Planet. Sci., 46* (Suppl.), A264.

Zurfluh, F.J., Hofmann, B.A., Gnos, E., Eggenberger, U. (2013). Sweating meteorites"-Water-soluble salts and temperature variation in ordinary chondrites and soil from the hot desert of Oman. *Meteorit. Planet. Sci., 48*(10), 1958-1980.
[http://dx.doi.org/10.1111/maps.12211]

CHAPTER 4

Meteorites of Northwest Africa: Morocco, Algeria, Mali, Mauritania

Lahcen Ouknine[1,2,*], Fouad Khiri[1,2,3] and Abderrahmane Ibhi[1,2]

[1] *Geoheritage and Geomaterials Laboratory, Ibn Zohr University, Agadir, Morocco*

[2] *University Museum of Meteorites, Ibn Zohr University, Agadir, Morocco*

[3] *University Museum of Meteorites, Regional Center of Trades of Education and Training, Inzegane, Agadir, Morocco*

Abstract: The number of meteorite finds in Northwest of Africa (NWA), *i.e.,* Morocco, Algeria, Mauritania and Mali have recorded a considerable increase since 1999. However, the classification of these meteorites is done by the Meteorite Nomenclature Committee of the "Meteoritical Society", only attributes 8% of the total finds of this region to their specific country of origin, and leaves 92% of them under the mere appellation "NWA" (Northwest Africa) followed by a number. This work attempts to contextualize the 5678 finds of NWA meteorites by defining the circumstances of the find of every sample, according to the new Categorization of Finds and the new Guidelines for Meteorite Nomenclature adopted by the Meteorite Nomenclature Committee. Thus, in addition to the 1180 official NWA meteorites whose countries of find are approved by the Nomenclature Committee, 3240 meteorites are assigned to 4 countries of North-West Africa, *i.e.,* 2994 samples (92%) to Morocco, 79 samples (2.5%) to Algeria, 34 samples (1.1%) to Mauritania and 12 samples (0.1%) to Mali. Nevertheless, the remaining NWA meteorites (1267 samples) have no information indicating the country of finding. After the adoption of the naming "NWA", we notice a remarkable decrease in the number of meteorites bearing the names of official places versus a considerable increase in the number of NWA meteorites. On the other hand, the statistical analysis of NWA meteorites reveals that the population includes rare specimens of great scientific value, making them highly desired by both scientists and collectors from all over the world. In general, this work results in the creation of a new database of meteorites stemming from the Northwest of Africa.

Keywords: Classification, New database, NWA meteorite finds, Renaiming, Statistics.

---

[*] **Corresponding author Lahcen Ouknine:** Geoheritage and Geomaterials Laboratory, Ibn Zohr University, Agadir, Morocco & University Museum of Meteorites, Ibn Zohr University, Agadir, Morocco;
E-mail: lahcen.ouknine@edu.uiz.ac.ma

Abderrahmane Ibhi, Giorgio S. Senesi, Lahcen Ouknine, Fouad Khiri (Eds.)

INTRODUCTION

In "cold" deserts like Antarctica, the collection of meteorites is prohibited (Schmitt, 2002), and is allowed exclusively for governmental scientific expeditions (especially from the United States and Japan), which possess all the technical capacities required to define the circumstances of meteorite finds (*e.g.*, the geographic coordinates of the harvest places). During the last 30 years, the specialized institutions of NASA recovered in this region more than 33000 meteorite samples, which were studied by more than 500 researchers all over the world (NASA, 2015). In contrast, in the "hot" desert region, especially in Northwest Africa, the collection and sale of meteorites are authorized without limitation in Morocco, Mauritania and Mali, whereas in Algeria, the search for meteorites and their sale has been strictly forbidden since 2004.

Often, meteorite harvest is made by nomads and sometimes by equipped meteorite hunters (Ibhi, 2014), and the acquisition is made by meteorite collectors. Since the end of 1999, the number of meteorites collected in Northwest Africa has known a considerable evolution (Grossman, 2000; Connolly *et al.*, 2007). However, the adopted legal and statutorily legislations (especially in Algeria) and lack of awareness on meteorites of more commercial interest often prevent nomads and meteorite hunters from revealing the harvest places. Consequently, the Meteorite Nomenclature Committee of the "Meteoritical Society", names most meteorites found in Northwest Africa with the acronym "NWA" (Grossman, 2000).

To justify this decision, the Meteorite Nomenclature Committee considered that «the reliability of locality information associated with these meteorites is difficult to assess because of the anonymity of all of the finders and most of the original sellers». Furthermore, the Committee justified this resolution by the fact that «these meteorites are all sold as Moroccan finds, but there are plausible reports that some were actually collected in Algeria». Thus, for approving and naming meteorites collected in Morocco and in surrounding countries, the Nomenclature Committee bases itself on the conditions contained in the Guidelines for meteorite nomenclature (Weisberg *et al.*, 2008), which requires that finds to be approved must be accompanied by documentation that includes the date of find, the name and address of finder, and reasonable proof of the location find (*e.g.*, a single photograph showing the meteorite *in situ*, a length-scale and an active GPS unit displaying the geographic coordinates). This provision deprives Northwest African countries of their meteoritic heritage (Larouci, 2014; Ouknine *et al.*, 2016). Fortunately, in February 2015, the Nomenclature Committee abolished the arrangement debated during its last meeting (Agee *et al.*, 2015), so assessing the

same treatment for meteorites native to Morocco and neighboring countries during their approval.

The present chapter aims to contextualize and document NWA meteorites collected until December 31st, 2014, by attempting to define the circumstances of the find of every sample to attribute it to the country of harvest and consequently propose a name for it.

COUNTRY OF ORIGIN OF NWA METEORITES

During its decision to assign the name "NWA" to meteorites recovered from Northwest of Africa, the Nomenclature Committee has not explicitly defined the countries that constitute this region. According to Ibhi (2014) and Chennaoui (2008), NWA meteorites include all samples found in Morocco, Algeria, Mauritania, Mali, Niger and Nigeria. Furthermore, websites that are interested in meteorites, such as "www.galactic.stone.com" allocate NWA meteorites to the Northwest Africa countries Morocco, Algeria, Mauritania, and Mali and other countries without mentioning which one, whereas the website "www.allmeteorite.com" asserts that these meteorites are stemming from Morocco, Algeria, Mauritania and Niger.

The data extrapolated from the official website (http ://www.lpi.usra.edu /mete or/met bull.php) of the Meteoritical Society reveal that the major part (92%) of the 8107 meteorites recovered in Northwest Africa is declared under the acronym NWA, whereas only 8% are classified under the name of official places of recovery in four countries, *i.e.,* Morocco, Algeria, Mauritania and Mali (Fig. 1). Most meteorites recovered in Algeria (83%) are attributed local names (5113 samples), whereas a high proportion of those recovered in the three other countries are declared under the acronym NWA. The results show that almost all meteorites (93%) found in Morocco are named NWA, and only 7% (74 samples) are classified with the name of the finding place. The number of meteorite finds is low in Mauritania, *i.e.,* 32 samples, of which 21 (66%) are assigned the name NWA, and only 15 finds are recorded for Mali, of which 11 samples with the acronym NWA. Noteworthy, three samples found in Libya are declared under the acronym NWA, *i.e.,* NWA 1237, NWA 1241, and NWA 1242, although this country is not situated in Northwest Africa.

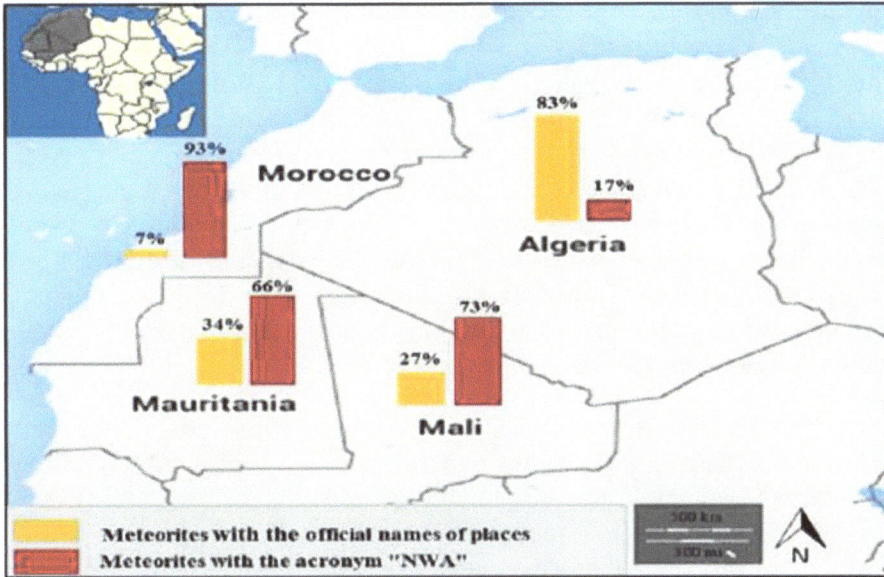

Fig. (1). The distribution of meteorites declared under the acronym NWA and those bearing the name of the official recovery place in the four Northwest African countries considered.

ALLOCATION OF NWA METEORITES TO THEIR COUNTRY OF ORIGIN

Circumstances of Discovery of NWA Meteorites

To clear up the circumstances of NWA meteorites, the exact specification of the place of discovery of every sample is considered of high relevance in this work. In particular, our search was based on the information contained in several databases on meteorites, *i.e.,* the Meteoritical Bulletin Database (http://www.lpi.usra.edu/meteor/metbull.php), the meteoritic database of NASA website (http://adsabs.harvard.edu/abstract_service.html), the database of the Natural History Museum of London (hhttp://www.nhm.ac.uk/nature-online/space/meteorites-dust//), and the database of the Meteorite Collection of the Department of Earth and Atmospheric Sciences of the University of Alberta (http://meteorite.museums.ualberta.ca//). Additional data were also collected from the confirmed sites of meteorite findings. These databases contain all information regarding each meteorite, *i.e.,* its name and abbreviation, its status (official or provisional), the year and place of its discovery, its type, the collected mass, and also its history, which retraces the circumstances of find and the last scientific publications on the meteorite in question.

On December 31, 2014, the number of NWA meteorites amounted to 8107, more than two-thirds of which, *i.e.,* 5687 meteorites, officially approved by the

Nomenclature Committee, 2414 samples (30%) in the provisional status, and 6 samples declared as terrestrial rocks. The data analysis of NWA meteorites officially classified and published in the Meteoritical Bulletin reveals that more than three-quarters of NWA finds (4507 samples) are declared without origin, whereas 1026 NWA meteorites are approved finds of Morocco, 124 of Algeria, 19 of Mauritania, and 11 of Mali. These numbers show that almost a quarter of NWA meteorites are allocated to the countries of finding, but without recognizing it officially with the attribution of the NWA acronym. Furthermore, 203 NWA meteorites feature the geographical coordinates of their find places, but do not bear the name of the "country of origin" instead of the acronym "NWA", which makes these meteorites anonymous.

In order to specify the circumstances of NWA meteorite finds, this research was based on: (i) the history and articles published on each official NWA meteorite; (ii) the authors responsible for the classification of the sample; and (iii) all other features on available on the meteorite (status, reclassification, pairing, *etc.*), especially the exact place of discovery (country, region, city or village). In particular, the definition of the discovery place was made according to the directives of the new Guidelines for Meteorite Nomenclature (Agee *et al.*, 2015). Furthermore, the proposed meteorite renaming followed the new Categorization of Finds and Falls adopted by the Nomenclature Committee (Agee *et al.*, 2015). However, the name NWA meteorite was kept for a meteorite that was found not to possess any track of its finding place.

This work allowed the formulation of a new database that assigns each NWA meteorite to one of the four Northwest African countries. On the basis of this new reallocation to the country of origin, the number of meteorites found in each of the four Northwest African countries has been reviewed. The results achieved are shown in Table **1**, and an extract of the database is presented in Table **2**.

Table 1. Number of official NWA meteorite finds before and after the reallocation to the country of origin.

Northwest African Country	Number of NWA Meteorites Officially Recognized	Number of Reallocated Meteorites	Total NWA Meteorites
Morocco	1026	+2994 (92.4%)	4020
Algeria	124	+79 (2.4%)	203
Mauritania	19	+34 (1.1%)	53
Mali	11	+12 (0.4%)	23
Morocco-Algeria border	0	+121(3.7%)	121

Table 2. Extract from the database developed after consideration of the circumstances of the find of various NWA meteorites.

Name (NomCom)	Abbreviation	Proposed Name*	Status	Year of Discovery	Place of Discovery*	Country of Discovery*	Latitude*	Longitude*	Number of Pieces Found*	Mass (g)	Type	Classification* (Author and Institution)	References*	MB
Northwest Africa 1	NWA 001	NWA – Alnif 001	Official	1999	AlnifZagoura	Maroc	31°7'N	5°11'W	1	1200	L6	A.Rubin et al., UCLA	Grossman, 2000	84
Northwest Africa 2	NWA 002	NWA – Rhessouane 001	Official	1999	RhessouaneTagounite	Maroc	29°55'N	5°35'W	many	234	EL6	S. V. Afanasiev, Vernad	Grossman, 2000	84
Northwest Africa 3	NWA 003	NWA – Rhessouane 002	Official	1999	RhessouaneTagounite	Maroc	29°55'N	5°35'W	1	120	H4	S. V. Afanasiev, Vernad	Grossman, 2000	84
Northwest Africa 10093	NWA 10093	NWA – Algérie 135	Official	2013	Algérie	Algérie	28°2'N	1°39'W	1	26 g	LL5	A.E. Rubin, UCLA	Agee et al., 2015	104†
Northwest Africa 10095	NWA 10095	NWA – NWA 1271	Official	2012	NWA	NWA	27°17'N	8°39'W	2	599 g	L4	A.E. Rubin, UCLA	Agee et al., 2015	104†
Northwest Africa 10096	NWA 10096	NWA – NWA 1272	Official	2013	NWA	NWA	27°17'N	8°39'W	1	159.1 g	L4	A.E. Rubin, UCLA	Agee et al., 2015	104†
Northwest Africa 10097	NWA 10097	NWA - Aoussred	Official	2013	Aoussred Dakhla	Maroc	22°33'	14°19'	many	186000	L6	A.E. Rubin and J. Utas, UCLA	Agee et al., 2015	104†
Northwest Africa 10099	NWA 10099	NWA - Ouarzazate 040	Official	2013	Ouarzazate	Maroc	30°53'N	6°55'W	1	3800	L6	A.E. Rubin and J. Utas, UCLA	Agee et al., 2015	104†

In particular, besides the 1026 NWA meteorites already declared by the Nomenclature Committee as found in Morocco, additional 2994 samples, which represent more than 92% of the reinstated official meteorites, have been discovered as stemming from this country, so reaching a total of 4020 meteorites (Table **1**). Among these, 2783 samples feature precise collection places that are distributed in the semi-arid areas located between the Southeast (Daraa and Tafilelt) and the Southwest region (Sagiua Elhamra and Wadi Eddahab) of the country. Actually, 150 meteorites from Morocco possess information on the region of finding, 2223 were collected in the perimeters of some cities, and the finding localities of the remaining 410 meteorites are exactly mentioned (village or douar) (Fig. **2**). However, the data available on 1237 samples do not provide the exact indications of the recovery place.

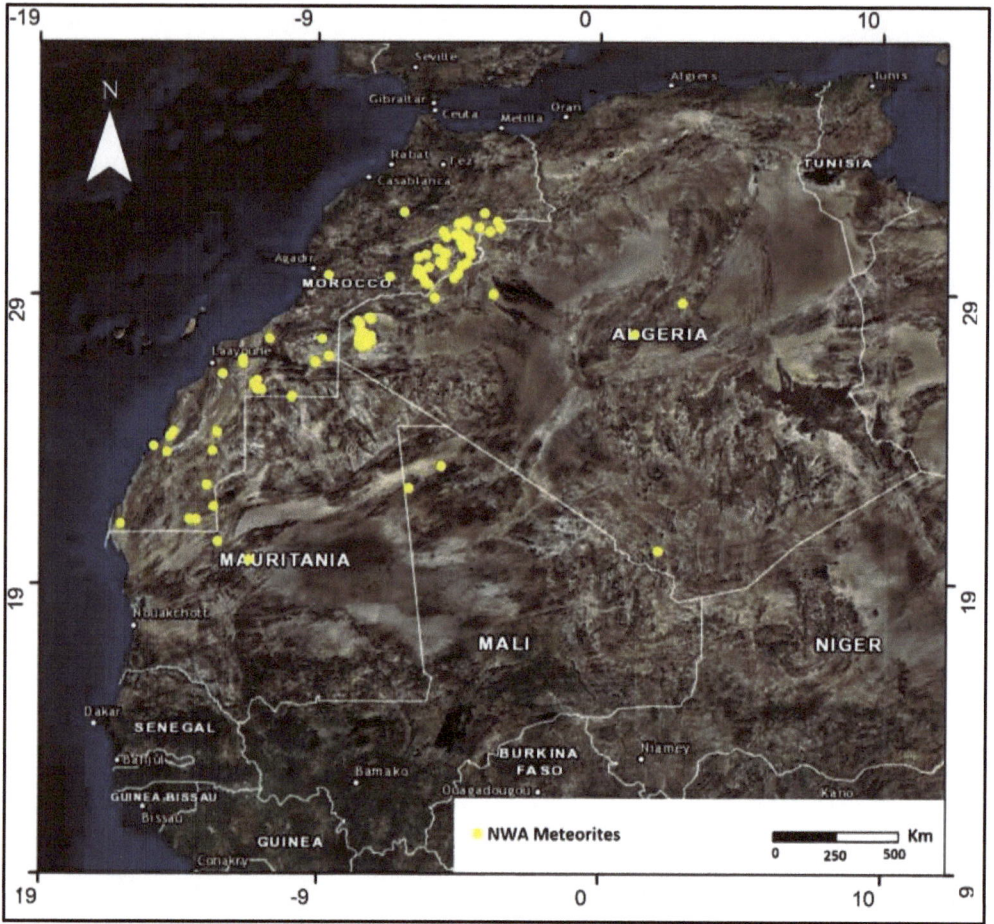

Fig. (2). Map showing the discovery places of NWA meteorites.

Regarding Algeria, 124 meteorites are already assigned by the Committee, and 79 samples, *i.e.,* 2.5% of the total number of samples reinstated in this work, are attributed to this country, for a total of 203 meteorites (Table **1**). In particular, the available information indicates that 120 samples were found in villages and 11 in city surroundings, 17 were recovered in various districts, especially that of Bechar, and 150 in the country without any indication of the region or the city of recovery.

Among the reallocated meteorites, 53 samples were recovered in Mauritania, of which 46 do not possess the information required to identify their exact places of find, 2 were collected in the region of Aatar Adrar, and 5 in the two cities of Nouadibou and Zouirat.

Until December 31, 2014, 23 meteorites were reported to be found in Mali, among which 12 samples have been reallocated after a confirmation of their finding places. In this work, the places of collection of 9 samples were defined, of which 5 were found in the cities of Gaw and Nouadibou and 4 in the localities of Guirat Laghnem, Bafana and Terhazza. However, 14 meteorites native to this country still lack the necessary data indicating their exact finding places.

Finally, of the 121 meteorites found in the border zone between Morocco and Algeria, those collected in Algeria and sold in Morocco were declared by the Nomenclature Committee as finds of Algeria, with the exception of 3 samples, *i.e.,* NWA 4448, NWA 4932 and NWA 4801, which were declared finds of Morocco, although collected in Algeria. Further 6 samples, *i.e.,* NWA 033, NWA 046, NWA 6204, NWA 2739, NWA 2017 and NWA 4590, were approved as finds of Algeria, but the circumstances of their finds would indicate that they are native of Morocco.

In conclusion, after the analysis of the circumstances of the find of each meteorite carrying the acronym NWA, 3240 NWA samples have been reallocated to four Northwest African countries. However, although recovered in Northwest Africa, 1267 (23%) remaining NWA meteorites do not possess information indicating the country of origin, as they were bought by collectors from various countries of the world, especially France, the United States and Germany. Thus, to avoid any confusion on place of their find, these meteorites continue to keep the acronym NWA. Thus, the decision of the Nomenclature Committee to assign the name NWA to almost all meteorites from Northwest Africa would need to be revised.

Renaiming Proposed for NWA Meteorites

In the world, the Meteorite Nomenclature Committee, which is part of the international, official, non-profit, scholarly organization named Meteoritical Society, is the only organ responsible for classifying and assigning any new meteorite a name that is the name of the place closest to its find. Since 1980, the meteorite nomenclature is adjourned according to the provisions contained in the Guidelines for Meteorite Nomenclature developed by the Nomenclature Committee that updates them continuously. After approval, the new names of finds are officially published in the Meteoritical Bulletin. The proposal advanced in this work is to rename NWA meteorites according to the specific and precise information acquired on the circumstances of find of each sample by applying revised terms for the Guide of Nomenclature (Agee, 2015) as listed in Table **3**.

Table 3. Renaming NWA meteorites according to new proposed terms of the Guidelines

Renaiming Proposed for NWA Meteorites	Terms of Guidelines Proposed to be Adopted for Renaiming Meteorites
Renomenclature of meteorites found in the mountains of the High and Middle Atlas, cities and villages.	**3.1 Geographic features:** «Acceptable names include physiographic features such as rivers, mountains, lakes, bays, capes and islands, political features such as towns, counties, states and provinces, and sites of human activity such as parks, mines, historical sites and railroad stations. »
Renomenclature of meteorites that only have data on the country, region or province of the find.	**3.1 Geographic features:** « The name of large geographic features such as continents, countries, states and provinces should be avoided if more specific names are available. »
Renomenclature of meteorites recovered in sparse places.	**3.3 Sparse place names** **Coincidental finds**. «Where several meteorites are found near the same locality, and alternative place names are not available, each separate meteorite shall bear the name of the locality followed by a parenthesized lower-case letter»
Renomenclature of meteorites recovered in some areas of Morocco, *i.e.,* Erfoud, Rissani and Laayoune.	**3.3 Sparse place names** **Dense collection areas**. «If particularly numerous recoveries are made in one region, new meteorites found within the designated region will be named by combining the prefix with the next available suffix.
Renomenclature of meteorites collected in localities mentioned by nomads, collectors and sellers.	**3.4 Meteorites of unknown or poorly known provenance.** **Withheld information**. «Where the source of a new meteorite cannot be determined due to the withholding of geographic information by a collector or other party, the name should be chosen to reflect the smallest geographic feature identifying the collection location with certainty. »
Renomenclature of meteorites found by sellers in places like Erfoud, Rissani, Zagora. *etc.*	**3.4 Meteorites of unknown or poorly known provenance.** **Transported meteorites**. «When the provenance of a new meteorite cannot be determined due to the lack of sufficient historical information, it should be named after the locality where it was first recognized. »

In this work, we referred to Section 3 (New Meteorite Names), Subsection 3.1 (Geographic features), to rename samples discovered in the mountains (the Martian NWA 1950 and the Lunar NWA 2200, for example) or in villages, and meteorites with information only on the country, the region or the province of discovery. To rename meteorites collected in scattered places (a frequent case in the Southeast of Morocco), we referred to Subsection 3.3 (Sparse placenames), paragraph (b) (Coincidental finds), which concerns the adopted renaming in the case when several samples are discovered near the same locality and names of alternative places are not available. This is the case, for example, of meteorites NWA 455, NWA 456, NWA 458, NWA 461, NWA 462, NWA 463, NWA 465, and NWA 773 that were found near the village of Ad Chira in the Southwest of Morocco (Russell *et al.*, 2003; Grossman and Zipfel, 2001). We also suggest renaming meteorites collected in places with a strong density of finds as it is the case of some areas in the Southeast of Morocco (Erfoud, Rissani and Zagora) with reference to the clauses of paragraph (c) (Dense collection areas) of the same Subsection 3.3. Furthermore, Subsection 3.4 (Meteorites of unknown or poorly known provenance), paragraph (a) (Withheld information) is used to rename samples collected by nomads, collectors and sellers of meteorites. Finally, the renaiming of dozens of meteorites found at dealers in certain cities, ie., Erfoud, Rissani and Temara in Morocco and Nouadhibou in Mauritania, is made referring to paragraph (b) (Transported meteorites) of Subsection 3.4. In particular, this clause is made for naming meteorites whose origin cannot be determined because of a lack of sufficient historical information, thus, these samples bear the names of the localities where they were recognized at first.

In this work, the use of the information extrapolated from the circumstances of discovery of NWA meteorites allowed to rename 77% of samples according to the clauses of the new Guide of Nomenclature of the "Meteoritical Society". However, 1267 meteorites (23%) will still keep the acronym NWA because of the absence of reliable information indicating their places of discovery.

Temporal Evolution of Total NWA Meteorite Finds and Total Located Meteorites in Four Northwest African Countries

In order to consolidate the results obtained after the reallocation of NWA meteorites to the countries of origin, the annual evolution of the total number of NWA finds has been compared with that of the located ones, *i.e.,* classified with the names of the official places, in the four Northwest African countries (Fig. **3**). This comparison also allows revealing the internal and external conditions that affect their collection. Indeed, 97.6% of NWA meteorites possess the year of discovery.

Fig. (**3**) shows that the number of NWA-named meteorite finds (ordinate on the left) is null before 1999, and oscillates between 1999 and 2014. In particular, after the adoption of the NWA acronym in the year 2000, the NWA meteorite finds showed an important progress reaching a number of 463 in 2004. This evolution can be explained by the development of the culture of "hunting for meteorites" encouraged by the enormous profits that pulled off the trade of meteorites. Successively, especially in the years 2007-2010, a remarkable decrease is recorded, which was feasibly due to both the political and security instability of the region with the appearance of armed groups in South Algeria, North Mali and East Mauritania and the partial decrease in the prices of "black rocks" because of the acronym NWA. Then, a new increase in NWA meteorites finds is recorded between 2011 and 2012, probably due to the progressive regularization of the security conditions in the Sahel. However, this period was followed by a new decrease in 2013 and, especially, 2014.

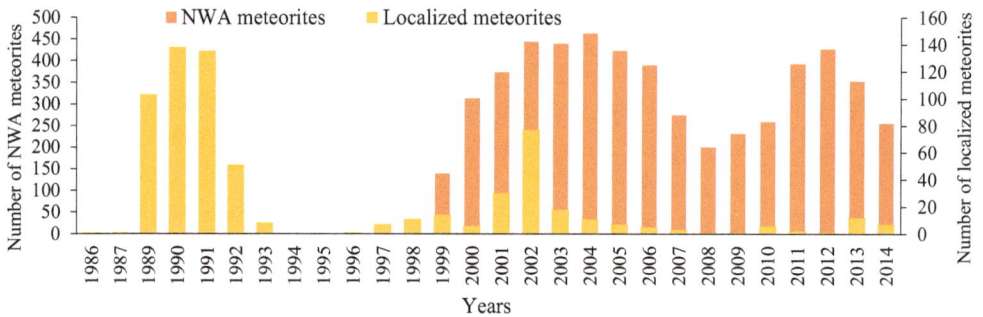

Fig. (3). Annual evolution of the number of NWA finds (ordinate on the left) and meteorites classified under the names of official places (ordinate on the right) in four Northwest African countries.

The first meteorite officially localized in the four Northwest Africa countries, and declared under the name of the official place of recovery was found in 1716 in "Kayes" in Mali and bears the name "Siratik". Also the number of localized meteorites ((Fig. **3**), ordinate on the right) oscillates with time from 1989 to 2014. During the years 1989-1991, a marked increase of Northwest African localized meteorites occurred, with 103 samples in 1989, 138 in 1990 and about 130 in 1991. Almost all meteorites declared during this period (99%) were found in Algeria (423 meteorites) and only 4 samples in Morocco. The discovery of dense collection areas (DCA) characterize this period, such as those in the district of Tamanghasset in Algeria where 316 meteorites were found in Reg El Acfer (for example, Acfer 001) and in the locality of Tanezrouft in the district of Adrar where 53 samples were collected (from Tanezrouft 001 to Tanezrouft 053) (Wlotzka, 1990, 1991, 1992a, b, 1993a, b). This period features the highest number of samples stemming from Northwest Africa and declared with their names of finding locations. This period was followed by a sudden decrease from 1992 to 1993, whereas in the years 1994, 1995 and 1996 no Northwest African meteorite find is reported to be localized. Starting in 1997 a new oscillating number of localized finds of Northwest African meteorite occurred, which reached a maximum in 2002 and then markedly decreased until 2014, with minima in 2007, 2008, 2011 and 2012. In the years 1997 to 1999 the number of declared finds were about 55, of which 36 samples (66%) collected in Morocco, especially in the Southwest DCA (Lahmada 001 to Lahmada 018). In particular, in the year 1999 only 6 meteorites (2% of finds) were declared under the names of the finding place, whereas 139 samples (98%) were published under the name NWA, so making them the first group of meteorites with this acronym. In the years from 2000 to 2004 a total of 107 meteorites were classified as localized, of which 87 of Algerian origin found especially in the DCAs of Tanezrouft (21 samples) and Acfer (66 samples). The increase of finds in Algeria during this period may be partially attributed to the formation of research teams on meteorites. In the period 2005-2014, relatively very few localized findings are reported, with 34 samples (79%) localized in Morocco, 5 samples in Mali and 3 of Mauritania. In particular, the 12 meteorites recognized in 2013 and the 7 in 2014 were all found in Morocco, which is, at least partially, due to the implementation of structures for meteorites search in this country. Noteworthy, the prohibition of the collection and sale of meteorites imposed in Algeria in 2004 would explain the decrease of the number of localized finds in all four Northwest African countries. However, this time was characterized by the dominance of NWA meteorites with the publication of 3987 samples that represent 99% of declared Northwest African meteorites.

In conclusion, the temporal evolution of meteorite finds in Northwest Africa shows that since the adoption of the name NWA in 2000, the classification is made extensively using this acronym (92%), and rarely under the exact place of finding (8%). In the following sections, a more detailed documentation and evaluation of the temporal evolution of the number of NWA meteorites finds and of those featuring the name of the finding localities are provided separately for each of the four Northwest African countries.

Temporal Evolution of NWA Meteorite Finds and Located Meteorites in Each of the Four Northwest African Countries

Morocco

The temporal evolution of the number of NWA finds and of meteorites classified under names of Moroccan localities appears very similar to those of the corresponding meteorites of Northwest Africa, showing various distinct phases (Fig. **4**). These samples represent almost 72.5% (4020 meteorites) of the total number found in Northwest Africa.

Morocco's first declared meteorite find bearing the acronym NWA is NWA 8144, which was discovered in 1993. The annual number of finds increased gradually since 2000 with the collection of 199 samples, reaching its peak in 2003 with 365 meteorites, which is especially due to the new finds in Southeast Morocco (Zagora, Erfoud and Rissani). After 2004, the number of finds decreased, reaching a minimum in 2010, then it increased again in 2011 and 2012, with a new minimum in 2013 and 2014.

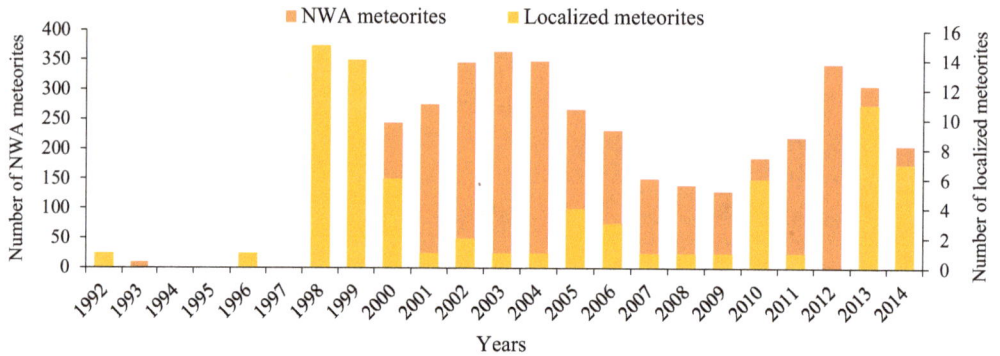

Fig. (4). Annual evolution of the number of meteorites finds in Morocco bearing the name NWA and the name of the finding locality.

The first meteorite bearing the name of the finding locality in Morocco is an iron meteorite weighing 79.9 kg and named Mrirte, which was found in 1937 in the neighborhood of the town of Mrirte in the Middle Atlas. Until 1997 the number of localized finds was very low, with the declaration of 6 meteorites only, while it increased in 1998 and 1999 ((Fig. **4**), ordinate on the right side), thanks to the discovery of a DCA in Lahmada (19 samples). After the year 2000, when the Nomenclature Committee hardened the conditions of the declaration of meteorites stemming from Morocco and neighboring countries (Grossman *et al.*, 2000), a marked reduction of finds in Morocco, with 6 meteorites found in 2000 and one to two samples declared every year between 2001 and 2004. However, an increase can be detected in 2005, 2006, 2010 and, especially more recently, *i.e.,* in 2013 and 2014.

Algeria

The NWA meteorites from Algeria comprise 203 samples which represent 3.7% of total NWA meteorites. After the adoption of the acronym "NWA", the annual discovery of NWA meteorites in Algeria increased from 6 samples found in 1999 to 20 in 2000, number that remained almost stable until 2003 (Fig. **5**). Then, the number of NWA meteorites fell abruptly to only 5 meteorites in 2004, because their collection in Algeria was strictly banned since January of that year. In the years 2005 and 2006, the number of NWA meteorites stemming from this country increased again, with 35 finds in 2005 and 33 in 2006, but actually, a major part of NWA meteorites found in Algeria after 2004 was sold in Morocco (Connolly *et al.*, 2006, 2007 and 2008), which explains the abnormal increase of finds in the next two years, 2005 and 2006, due to the time required to sell them outside Algeria. The number of NWA meteorite finds declined again to 13 samples in 2007 and below 5 every year until 2014.

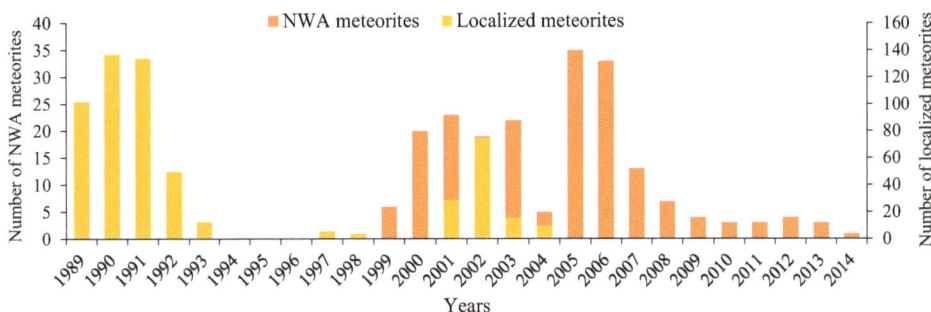

Fig. (5). Annual evolution of the number of meteorite finds in Algeria bearing the name NWA and the name of the finding place.

The first official meteorite found in Algeria was recovered in the region of Adrar in 1864, is of metallic type (Iron, IIIAB), weighs 510 kg and is called "Tamentit" (Grady, 2000; Kolbitz, 2006). Between 1864 and 1969, 19 meteorites were declared as finds of Algeria, with 8 samples found in the region of Adrar. In the period 1989 to 1992, 428 meteorites were declared under the names of their discovery locality, which represent 71% of all finds located in Algeria. The huge number of finds during these four years was due to the discovery of 316 meteorites in the DCA of Tamanghasset (Reg El Acfer) and 53 samples in Tanezrouft (Adrar) (Wlotzka, 1990, 1991, 1992 and 1993). In the period 1993 to 1998, the number of finds located in Algeria fell to 13 meteorites in 1993 and only 4 in 1998, which can be due to the insecurity which reigned in South Algeria during these years because of the rise of armed groups. After two years (1999 and 2000) that recorded no new find of localized meteorites in Algeria, in the years between 2001 and 2004, an abrupt increase occurred, with 29 samples found in 2001, a peak of 75 finds in 2002, and a total of 130 classified samples that represent 22% of all localized finds. This increase can be ascribed to the discovery of new types of meteorites in the DCA of Tanezrouft (36 meteorites, from Tnz 54 to Tnz 89) and that of Reg El Acfer (74 samples, from Acfer 321 to Acfer 401), which was favored by the improvement of security conditions in South Algeria. However, in the years 2003 and 2004, the frequency of located finds decreased to 16 and 10 samples, respectively. From January 2004, the Algerian government made the decision to forbid the collection of meteorites on its territory, which explains the absence of finds declared under the local names after this year.

In conclusion, the number of meteorites declared officially under the names of Algerian localities (602 samples) exceeds widely the number of meteorites classified under the general name NWA (203 samples). The ban on the harvest of «black rocks» in this country in 2004 caused the stop of the declarations of meteorites under the names of the finding locality and the progressive reduction of the number of meteorites classified under the NWA acronym to arrive to a single meteorite declared in the year 2014.

Mauritania

Of all NWA meteorites, 53 samples, *i.e.,* 1% were collected in Mauritania (Fig. **6**). After one year from the adoption of the NWA appellation, the classification of NWA meteorites native of Mauritania started in 2001 with the declaration of the NWA 1769, which is an achondrite howardite type (Russell *et al.*, 2004), and the NWA 3163 of lunar type (Connolly *et al.*, 2006). The NWA Finds increased to 8 meteorites in 2002 but declined to only 2 samples in 2003.

The period between 2004 and 2009 is characterized by no find, with two exceptions in 2005 and 2008 when two meteorites were collected. This break may be ascribed to the lack of rising citizens awareness on the scientific importance and added socioeconomic value of meteorites, and the insecurity which reigned in the east of the country. However, in the years between 2010 and 2014, the number of NWA meteorites stemming from Mauritania increased with the recovery of 2 meteorites in 2010, 16 in 2012, 4 in 2013 and 17 in 2014, for a total of 39 samples, that represent 75% of NWA meteorites collected in Mauritania. This increase may be attributed to dealers who acquired these rocks without revealing the exact places of finds.

In Mauritania, the total number of meteorites named under the place of find is 15 spread on seven regions of the country and collected along 94 years (Fig. **6**). The first located meteorite was collected in the region of Adrar in 1920 and is named "Chinguetti", which is a type of mesosiderite B1 (Grady, 2000; Kolbitz, 2006) and weighs more than 5 kg. The located finds in this country are spaced out in time and do not exceed one-three meteorites at most in a year. However, since 2001, a certain continuity of meteorite recovery has been recorded in Mauritania with their declaration either under the name of the place of find or under the general acronym NWA. Nevertheless, the dominance of the number of NWA meteorites on that of located meteorites reflects the lack of consciousness of nomads and local collectors on the importance to reveal the exact places of find to better protect their meteoritic heritage.

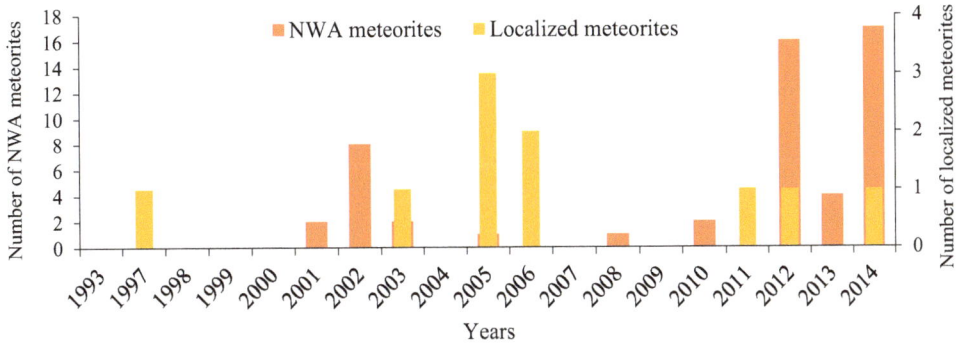

Fig. (6). Annual evolution of the number of meteorite finds in Mauritania bearing the name NWA and the name of the finding place.

Mali

Despite its large surface that exceeds a million square km (1104964 km^2), of which 90% is desert (World Bank, 2012), the total NWA meteorites native of Mali is low and does not exceed 19 samples. Nevertheless, Mali is the only one of the 4 countries of Northwest Africa where meteorites are officially declared, and none is in provisional status.

After the adoption of the appellation NWA in 2000, 2 meteorites found in the municipality of Terhazza near the city of Tombouctou were called NWA 512 and NWA 513, although the geographical coordinates of their harvest places are known (Grossman and Zippfel, 2006). Then, a break of a collection of NWA meteorites in Mali occurred between 2000 and 2007, which can be attributable, at least partially, to the effects of political and security instability. However, in the period between 2008 and 2014, the collection of NWA meteorites records 90% in Mali, with one peak in the year 2010 with 6 meteorites (Fig. 7).

Only 7 meteorites from Mali bear the name of the collection place, and are spread over 3 centuries, with the first one called "Siratik" found in 1716, "Gao," "N'Goureyma" and "Trifir" found in the 20th and "Chergach," "Dar El Kahal" and "Taoudenni" in the 21st century.

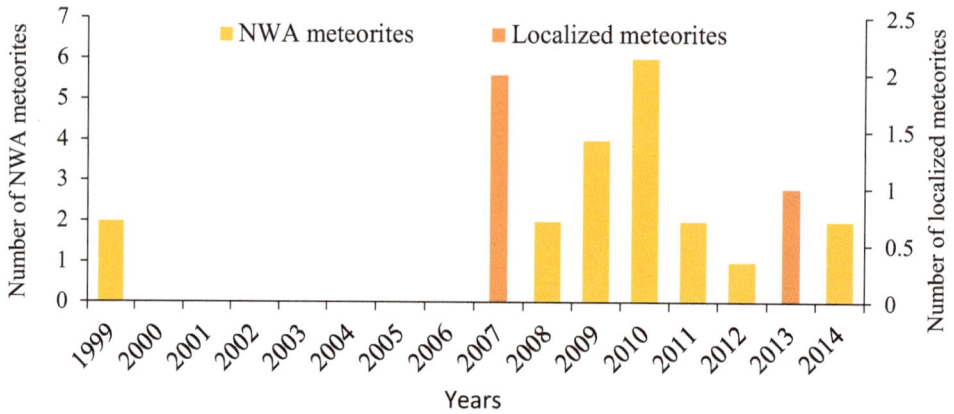

Fig. (7). Annual evolution of the number of meteorite finds in Mali bearing the name NWA and the name of the finding place.

In conclusion, after the reallocation attempted in this work of the 3240 NWA meteorites to the 4 Northwest African countries, 1267 samples still remain without affiliation because of the lack of sufficient indications on the circumstances of their finds. Actually, the number of meteorites without pedigree has increased gradually since the adoption of the naming NWA, particularly between 2004 and 2007 with only 464 samples collected, which was possibly due to the ban on meteorites collection in Algeria in 2004. However, the number of NWA meteorites without precise origin has decreased yearly since 2006 to attain 29 samples in 2014.

CLASSIFICATION OF NWA METEORITES

NWA meteorites are mostly stony meteorites (97%) with very few stony-iron and iron meteorites of identical amount (1.5% each) (Table **4**). Of stony meteorites, 78% are chondrites, less than the percentage recorded at the world level (Guignard, 2011) that amount to 90%, and 22% are achondrites. More than half of achondrites are of the HED type (56%) followed by ureilites (14%) and 12% of Primitive Achondrites. NWA meteorites also include one third (33.65%) of world mesosiderites (217 samples).

More than three-quarters of NWA meteorites are Chondrites (4287 samples), of which 86% are ordinary Chondrites, and the remaining are Carbonaceous Chondrites (10%), Enstatites (2%) and Chondrites of type Rumuruti (2%). Half of the ordinary chondrites are of type L, and the other half is distributed between types H (33%) and LL (19%). Of the Carbonaceous Chondrites, the most known is the NWA 5958, which was bought from a meteorite hunter from Tagounit in Southeast Morocco in 2009, and is classified as an Ungrouped Carbonaceous Chondrite as the unique primitive carbonaceous chondrite. This meteorite is the only find of the hot deserts that contains tracks of extraterrestrial amino acids (Osawa, 2014), and features extreme isotopic signatures that provide unique information about the forming of presolar dust grains (Bunch *et al.*, 2011; Nittler *et al.*, 2012).

Table 4. Abundance of NWA meteorites.

Types				Abundance (%)	
Stony meteorites	Chondrites	Carbonaceous		7.80	78
		Ordinary	L	31.40	
			LL	13.80	
			H	21.20	
		Rumuruti		1.69	
		Enstatite Chondrites		1.88	
		Ungrouped Chondrites		0.23	
	Achondrites	Primitive achondrites		2.64	22
		Enstatite achondrites		0.21	
		Aubrites		0.20	
		Angrites		0.21	
		HED		12.32	
		Ureilites		3.08	
		Lunar meteorites		1.54	
		Martian meteorites		1.54	
		Ungrouped Achondrites		0.00	
Stony–iron meteorites		Pallasites		0.19	1.49
		Mesosiderites		1.30	
Iron meteorites		Irons		1.33	1.33
Total				100	100

The NWA 7325 discovered in 2012 was attributed to the Achondrite ungrouped type and considered as the first meteorite native of the crust of Mercury (Irving and Kuehner, 2013), but subsequent studies (Goodrich, 2014; Kita, 2014) indicated that this meteorite is of Ureilite type stemming from an asteroid. Of NWA Achondrites 89 are Lunar meteorites, 15 of which were classified in 2014 with an estimated mass of 49 kg. The NWA 500 is the biggest Lunar meteorite (11.53 kg) discovered on Earth originated from an unexplored part of the Moon and represented the only available big gabbro of Lunar land (Connolly *et al.*, 2007). Of Martian meteorites, 82 samples were discovered in Northwest Africa (7% of Achondrites) and included the 7034 NWA type basalt breccia. This meteorite is called "Black Beauty" instead of the proper name that should be "Rabt Sbayta" (Larouci, 2014), and is different from other Martian meteorites by its relatively rich content in water (Agee *et al.*, 2013). Furthermore, NWA 2737 (Barrat *et al.*, 2000) and NWA 8694 (Hewins *et al.*, 2014) collected in Morocco are two meteorites of rare type (Chassignite) and great scientific value.

METEORITES FROM NORTHWEST AFRICA: A DISSIPATED HERITAGE

The number of officially declared meteorites for Northwest African countries does not exceed 8% of the total number of meteorites collected in this region, whereas major part of them (92%) remain declared under the general name NWA.

Established databases show that NWA meteorites represent 16% of the total number of world finds (a meteorite out of six is NWA). Although the surface of Northwest African countries represents only 3.6% of the Earth surface, one sixth of world meteorites are without affiliation and scattered in 4 countries (Morocco, Algeria, Mali, and Mauritania). The lack of knowledge of meteorites origins represents a loss of significant information which turns in a negative impact on an important part of the geologic heritages of Northwest African countries. The use of the generic acronym NWA for meteorites that are stemming from this region also implies negative scientific repercussions as it does not allow the identification of pairing samples. As an example, NWA 8005 and NWA 8006 are similar to NWA 7934 (Ruzicka *et al.*, 2015). The considerations above confirm that NWA samples are treated by researchers just as meteorites without considering their origins with no interest in the scientific contribution that they would deserve.

Moreover, the consideration of NWA meteorites as anonymous meteorites without pedigree has negative repercussions on the conservation of meteoritic heritage of the concerned countries. This heritage is getting lost without considering its inestimable scientific value. The French company "the memory of the earth" exported 24 tons of meteorites from the Southeast Morocco (Nicot, 2008) between 1999 and 2003, whereas for the same period, the mass of meteorites declared for Morocco does not exceed 1.86 t, which means 13 times less than the actual recovered amount only for this company.

Quoted examples prove that the quantities provided to scientific laboratories to widen the research on meteorites are too low with respect to the quantities that circulate in the meteorite world markets. Furthermore, the naming NWA causes the loss of visibility that may enjoy the countries of Northwest Africa in the case meteorites from these countries would assume the local names (cultural and ecotourism losses), and also the loss of socioeconomic benefits due to the devaluation of samples collected from this region, in spite of their possible rarity and important scientific value.

Finally, especially the hot zones of Northwest African countries that have recorded many finds suffer very much from their socioeconomic conditions, which are characterized by moderate literacy rates (31.1% in Mali, 58% in Mauritania, 67% in Morocco and 72.6% in Algeria), and by alarming poverty levels, especially in Mali and Mauritania (43.6% of the population lives under the poverty level in Mali, 42% in Mauritania, 8.9% in Morocco and 5.9% in Algeria) (World Bank, 2015). At the institutional level, these countries lack specialized structures (equipped scientific laboratories, dedicated museums) that can take care of the study and conservation of the geologic heritage, and in particular of meteoritic heritage (Darbali and Ibhi, 2010). Certainly, the conservation of geologic heritage is a concern of the geologic community, which is realized through the collections in museums (De Wever, 2008). In this perspective, the implementation of the first Museum of Meteorites in Africa and the Middle East at the University of Ibn Zohr in Morocco constitutes the first step to protect and highlight this non-renewable heritage (Ibhi *et al.*, 2016; Ibhi, 2016).

In our opinion, in such socioeconomic conditions the decision of the Meteorite Nomenclature Committee seems unreasonable because it is established on prerequisites that are supposed to be available wherever in the world without considering socioeconomic disparities between regions and countries.

CONCLUSION

The temporal evolution of the number of meteorites bearing the NWA acronym and that of those bearing the names of the finding place in the four Northwest African countries is characterized by two distinct periods. Before the decision of the Meteorite Nomenclature Committee to allocate the acronym NWA to meteorites resulting from this region in 2000, the number of meteorites finds registered a considerable increase, especially in Algeria. After the adoption of the acronym NWA in 2000, a remarkable decrease occurred in the number of meteorites bearing the names of official places (8%), and a considerable increase occurred in the number of NWA-named meteorites (92%). Furthermore, the new legal and statutorily legislations, especially in Algeria, and the lack of interest in meteorites of commercial interest, as well as the political instability and security in the Sahel region, have accelerated the recognition of meteorites stemming from Northwest Africa under the acronym NWA.

The results of this work, which aimed to contribute to the circumstances of finds of NWA meteorites collected until December 31st, 2014, show that most of NWA meteorites stem from Morocco and that only 23% of them have no information about the country of origin. The exploitation of the new database proposed in this work, after the study of the circumstances of discoveries of all NWA meteorites, allows to rename 77% of NWA samples according to the clauses of the new Guide of Nomenclature of the Meteoritical Society. However, 1267 meteorite samples (23%) would maintain the acronym "NWA" because of the absence of reliable information indicating the discovery locality. This proves that the decision of the Nomenclature Committee on the allocation of the naming NWA to almost all meteorites of Northwest Africa would deserve an accurate revision.

The exploitation of the new database for NWA meteorites proposed in this work, as an information system, is expected to allow an efficient management of data collected on NWA meteorites and a reliable statistical analysis of their features. Finally, the proposed renaiming of NWA meteorites, which is in line with the directives of the new Guidelines for Meteorite Nomenclature adopted by the Meteorite Nomenclature Committee in February 2022, is reasonably expected to increase the value of the inestimable scientific and socioeconomic priceless meteoritic heritage of Morocco and other Northwest African countries.

REFERENCES

Agee, C.B., Wilson, N.V., McCubbin, F.M., Ziegler, K., Polyak, V.J., Sharp, Z.D., Asmerom, Y., Nunn, M.H., Shaheen, R., Thiemens, M.H., Steele, A., Fogel, M.L., Bowden, R., Glamoclija, M., Zhang, Z., Elardo, S.M. (2013). Unique meteorite from early Amazonian Mars: water-rich basaltic breccia Northwest Africa 7034. *Science, 339*(6121), 780-785.
[http://dx.doi.org/10.1126/science.1228858] [PMID: 23287721]

Agee, C., Bullock, E., Bouvier, A., Dunn, T., Gattacceca, J., Grossman, J., Herd, C., Ireland, T., Metzler, K., Mikouchi, T., Ruzicka, A., Smith, C., Welten, K., Welzenbach, L. (2015). Categorization of Finds and Falls. *Published in Meteoritical Society website.* Available at: http://www.lpi.usra.edu/meteor/docs/falls-finds.pdf(Accessed on: 2015).

Agee, C., Bullock, E., Bouvier, A., Dunn, T., Gattacceca, J., Grossman, J., Herd, C., Ireland, T., Metzler, K., Mikouchi, T., Ruzicka, A., Smith, C., Welten, K., Welzenbach, L. (2015). Guidelines for meteorite nomenclature. *Published in Meteoritical Societywebsite.* Available at: http://meteoriticalsociety.org/?page_id=59(Accessed on: 2015).

Aoudjehane, C.H. (2008). Les météorites du Maroc : Une richesse scientifique et un patrimoine à préserver. Actes de la session plénière solennelle de l'académie Hassan II des Sciences et Techniques.20-22 février 2008, Rabat, Maroc. pp. 111-131.

Aoudjehane, C.H. (2013). On meteorites from Morocco and the NWA meteorite nomenclature *76th Annual Meteoritical Society Meeting,* 5347.

Badyukuv, D. (2002). Classification and mineralogy of meteorite NWA 1237. The Meteoritical Bulletin, 86.

Meteorit. Planet. Sci., 37, 157-184.

Bland, P.A., Smith, T.B., Jull, A.J.T., Berry, F.J., Bevan, A.W.R., Cloudt, S., Pillinger, C.T. (1996). The flux of meteorites to the Earth over the last 50 000 years. *Mon. Not. R. Astron. Soc., 283*(2), 551-565.
[http://dx.doi.org/10.1093/mnras/283.2.551]

Bunch, T., Wittke, J. (2002). Classification and mineralogy of meteorite NWA 1242. The Meteoritical Bulletin, 86. *Meteorit. Planet. Sci., 37*, 157-184.

Connolly, H.C., Jr, Zipfel, J., Grossman, J.N., Folco, L., Smith, C., Jones, R.H., Righter, K., Zolensky, M., Russell, S.S., Benedix, G.K., Yamaguchi, A., Cohen, B.A. (2006). The Meteoritical Bulletin, No. 90, 2006 September. *Meteorit. Planet. Sci., 41*(9), 1383-1418.
[http://dx.doi.org/10.1111/j.1945-5100.2006.tb00529.x]

Connolly, H.C., Jr, Zipfel, J., Folco, L., Smith, C., Jones, R.H., Benedix, G., Righter, K., Yamaguchi, A., Aoudjehane, H.C., Grossman, J.N. (2007). The Meteoritical Bulletin, No. 91, 2007 March. *Meteorit. Planet. Sci., 42*(3), 413-466.
[http://dx.doi.org/10.1111/j.1945-5100.2007.tb00242.x]

Connolly, H.C., Jr, Smith, C., Benedix, G., Folco, L., Righter, K., Zipfel, J., Yamaguchi, A., Aoudjehane, H.C., Aoudjehane, H.C., Grossman, J.N. (2007). The Meteoritical Bulletin, No. 92, 2007 September. *Meteorit. Planet. Sci., 42*(9), 1647-1694.
[http://dx.doi.org/10.1111/j.1945-5100.2007.tb00596.x]

Connolly, H.C., Jr, Smith, C., Benedix, G., Folco, L., Righter, K., Zipfel, J., Yamaguchi, A., Aoudjehane, H.C., Aoudjehane, H.C., Grossman, J.N. (2008). The Meteoritical Bulletin, No. 93, 2008 March. *Meteorit. Planet. Sci., 43*(3), 571-632.
[http://dx.doi.org/10.1111/j.1945-5100.2008.tb00673.x]

Darbali, M., Ibhi, A. (2010). Exposition des météorites au Musée du Patrimoine d'Agadir : Exemple de la préservation et de la valorisation du patrimoine géologique.3eme Rencontre sur la Valorisation et la Préservation du Patrimoine Paléontologique. Agadir. Maroc: 6-8.

De Wever, P. (2008). Actes de la session plénière solennelle de l'académie Hassan II des Sciences et Techniques. Management of the Geological Heritage. février. Rabat, Maroc: 57-62.

Goodrich, C.A., Kita, N.T., Nakashima, D. (2014). Petrology of the NWA 7325 Ungrouped achondrite meteorite from mercury, the ureilite parent body, or a previously unsampled asteroid? 45th lunar and planetary science conference held 17-21 March, at The Woodlands, Texas. LPI Contribution. *1777*, 1246.

Grady, M. M. (2000). Catalogue of Meteorites, 5th Edition, Cambridge University Press, Edinburgh, UK.

Grossman, J. (2000). The Meteoritical Bulletin, No. 84, 2000 August. *Meteorit. Planet. Sci., 35*(S5), A199-A225.
[http://dx.doi.org/10.1111/j.1945-5100.2000.tb01797.x]

Grossman, J.N., Zipfel, J. (2001). The Meteoritical Bulletin, No. 85, 2001 September. *Meteorit. Planet. Sci., 36*(S9), A293-A322.
[http://dx.doi.org/10.1111/j.1945-5100.2001.tb01542.x]

Halliday, I., Griffin, A.A., Blackwell, A.T. (1996). Detailed data for 259 fireballs from the Canadian camera network and inferences concerning the influx of large meteoroids. *Meteorit. Planet. Sci., 31*(2), 185-217.
[http://dx.doi.org/10.1111/j.1945-5100.1996.tb02014.x]

Ibhi, A. (2014). Morocco meteorite falls and finds: some statistics. International Letters of Chemistry. *Physics and Astronomy, 1*, 18-24.

Ibhi, A. (2016). Musées et vulgarisation scientifique : cas du Musée Universitaire de Météorites. Premières Journées Nationales de la Recherche Scientifique et de l'Innovation.30 et 31 mai 2016, Agadir, Maroc..

Ibhi, A. Douzi, H., Khiri, F. & Ouknine, L. (2016). Musée Universitaire de Météorites, un projet participative. Premier Colloque International, la Muséologie dans le champ de la Culture Scientifique et Technique. mai 2016, Rabat, Maroc. 16-18.

Irving, A., Kuehner, S. (2013). Classification and mineralogy of meteorite NWA 7325. The Meteoritical Bulletin, 101. *Meteorit. Planet. Sci.,* 47.

Kita, N.T., Sanborn, M.E., Yin, Q.Z., Nakashima, D., Goodrich, C.A. (2014). The NWA 7325 Ungrouped achondrite-possible link to ureilites? oxygen and chromium isotopes and trace element abundances. *In: Lunar and Planetary Science Conference., 45,* 1455.

Koblitz, J. (2006). Metabase, version 7.1 (CD ROM).

Larouci, N., Aoudjehane, H.C., Jambon, A. (2014). Methodology of studying Moroccan meteorites: alternative solution to "NWA" nomenclature. *77th Annual Meteoritical Society Meeting,* 5349.

Metzler, K., Bischoff, A., Greenwood, R.C., Palme, H., Gellissen, M., Hopp, J., Franchi, I.A., Trieloff, M. (2011). The L3-6 chondritic regolith breccia Northwest Africa (NWA) 869: (I) Petrology, chemistry, oxygen isotopes, and Ar-Ar age determinations. *Meteorit. Planet. Sci., 46*(5), 652-680. [http://dx.doi.org/10.1111/j.1945-5100.2011.01181.x]

NASA website Available at: http://wwwcurator.jsc.nasa.gov/antmet/index.cfm(accessed on: 2015).

Nicot, F. (2008). Chercheurs de pépites célestes, in Sciences et Vie junior, 226.

Ouknine, L., Khiri, F., Ibhi, A. (2016). Study of the circumstances of meteorites "Northwest Africa" finds: contribution to an appropriate renomenclature. *79th Annual Meeting of the Meteoritical Society,* 6567.

Rubin, A. (2006). Classification and mineralogy of meteorite NWA 869. The Meteoritical Bulletin, 90. *Meteorit. Planet. Sci., 41*(9), 1383-1418.

Russell, S.S., Folco, L., Grady, M.M., Zolensky, M.E., Jones, R., Righter, K., Zipfel, J., Grossman, J.N. (2004). The Meteoritical Bulletin, No. 88, 2004 July. *Meteorit. Planet. Sci., 39*(S8) (Suppl.), A215-A272. [http://dx.doi.org/10.1111/j.1945-5100.2004.tb00357.x]

Russell, S.S., Zipfel, J., Folco, L., Jones, R., Grady, M.M., McCoy, T., Grossman, J.N. (2003). The Meteoritical Bulletin 85. *Meteorit. Planet. Sci., 36,* 293-322.

Ruzicka, A., Grossman, J., Bouvier, A., Herd, C., Agee, C. B. (2015). The Meteoritical Bulletin 102.

Ruzicka, A., Grossman, J., Bouvier, A., Herd, C., Agee, C.B. (2015). The Meteoritical Bulletin 102. *Meteorit. Planet. Sci., 50,* 1497-1662.

The Meteoritical Society website Available at: http://www.lpi.usra.edu/meteor/docs/ mb102.pdf(Accessed on: 2015).

Weisberg, M.K., Smith, C., Benedix, G., Folco, L., Righter, K., Zipfel, J., Yamaguchi, A., Aoudjehane, H.C. (2008). Guidelines for meteorite nomenclature. Published in Meteoritical Bulletin, 94. *Meteorit. Planet. Sci., 43,* 1551-1588. [http://dx.doi.org/10.1111/j.1945-5100.2008.tb01027.x]

Welten, K., Caffee, M.W., Franke, L., Timothy JULL, A.J., LeCLERC, M.D., Metzler, K., Ott, U. (2011). The L3-6 chondritic regolith breccia Northwest Africa (NWA) 869: (II) Noble gases and cosmogenic radionuclides. *Meteorit. Planet. Sci., 46*(7), 970-988. [http://dx.doi.org/10.1111/j.1945-5100.2011.01204.x]

Wlotzka, F., Kurz, M. (2002). Classification and mineralogy of meteorite NWA 1241, The Meteoritical Bulletin, 86. *Meteorit. Planet. Sci., 37,* 157-184.

Wlotzka, F. (1990). The Meteoritical Bulletin. *Meteoritics, 25*(3), 237-239. [http://dx.doi.org/10.1111/j.1945-5100.1990.tb01004.x]

Wlotzka, F. (1991). The Meteoritical Bulletin, No. 71. *Meteoritics, 26*(3), 255-262. [http://dx.doi.org/10.1111/j.1945-5100.1991.tb01047.x]

Wlotzka, F. (1992). The Meteoritical Bulletin, No. 72. *Meteoritics, 27*(1), 109-117. [http://dx.doi.org/10.1111/j.1945-5100.1992.tb01068.x]

Wlotzka, F. (1992). The Meteoritical Bulletin 73. *Meteoritics, 27,* 444-483.

Wlotzka, F. (1993). The Meteoritical Bulletin, No. 74, 1993 March. *Meteoritics, 28*(1), 146-153. [http://dx.doi.org/10.1111/j.1945-5100.1993.tb00262.x]

Wlotzka, F. (1993). The Meteoritical Bulletin, No. 75, 1993 December. *Meteoritics, 28*(5), 692-703. [http://dx.doi.org/10.1111/j.1945-5100.1993.tb00641.x]

CHAPTER 5

Meteorites of Morocco: Falls and Finds

Abderrahmane Ibhi[1,2,*], **Fouad Khiri**[1,2], **Giorgio S. Senesi**[3] and **Hassan Nachit**[2]

[1] *University Museum of Meteorites, Ibn Zohr University, Agadir, Morocco*

[2] *Geoheritage and Geomaterials Laboratory, Ibn Zohr University, Agadir, Morocco*

[3] *CNR, Institute for Plasma Science and Technology (ISTP), Bari seat, 70126 Bari, Italy*

Abstract: Since the first recorded discovery of a meteorite in 1937 near the Mrirt village, a total of 1747 authenticated meteorites have been recovered in Morocco. The collection includes 21 meteorites from authenticated observed falls, which comprise five different types of meteorites, *i.e.*, 17 ordinary chondrites including four LL6, one EH7, one EH5, one H3-5, one H3-6, four H5, one CH, two L6 and two L4, one carbonaceous chondrite, one eucrite unbrecciated, one shergottite basaltic achondrites and one aubrite. The meteorite fall recovery rate in Morocco during the past 88 years (from 1932 to 2020) on a surface area of 712. 5 km^2 is relatively low, with approximately four falls recorded every 10 years, equivalent to 0.4 falls per year per 2.11 km^2. A total of 1747 distinct and authenticated meteorite finds with a total mass of 6.175 kg have been described, which include 1674 stones (442 achondrites and 1232 chondrites), 37 iron, 24 stony-iron and 12 ungrouped meteorites. The rate of recovery of meteorites (falls + finds) in Morocco exceeds that of most other countries of similar size and climatic conditions. More than 95% of documented meteorites from Morocco, including many rare types, have been recovered from the Eastern Moroccan Sahara, which has been proven to be one of the most prolific world areas for meteorite finds.

Keywords: Classification, Meteorite falls and finds, Morocco, Statistics.

INTRODUCTION

Morocco is one of the richest countries in the world in terms of meteorite discoveries because: (i) meteorites are easily revealed by their color in the desert context; (ii) the dry climate helps to preserve them much better than a humid climate; (iii) the political stability of Morocco secures and facilitates the search

---

* **Corresponding author Abderrahmane Ibhi:** University Museum of Meteorites, Ibn Zohr University, Agadir, Morocco & Geoheritage and Geomaterials Laboratory, Ibn Zohr University, Agadir, Morocco; E-mail: a.ibhi@uiz.ac.ma

Abderrahmane Ibhi, Giorgio S. Senesi, Lahcen Ouknine, Fouad Khiri (Eds.)

for meteorites. On February, 2016, with the aim of popularizing the science of meteorites, the Ibn Zohr University Museum of Meteorites was inaugurated in the heart of the university complex to host a permanent exhibition of meteorites that crashed in Morocco. Actually the plan of founding a meteorite museum was not new, but emerged from the propitious maturation of reflections elaborated over the past twenty years. Indeed, since 2006, the Ibn Zohr Scientific and Cultural Club of Agadir, in collaboration with Ibn Zohr University and the National Network for the Promotion and Diffusion of Scientific and Technical Culture of CNRST (Rabat), has organized several temporary exhibitions of meteorites and associated rocks in different cities of Morocco, *i.e.*, Agadir and Mir Left in 2006; Rabat and Guelmim in 2008; Casablanca in 2010; Tata in 2013 and 2015; *etc.*).

The Museum consists of two rooms; the one is a meteorite exhibition room that presents a collection of meteorites of great beauty that includes more than 150 meteorites, of which 50 impactites (tektites, shatter-cones, impact breccias, *etc.*) with rock engravings (Fig. **1**); the other is a projection room dedicated to conferences, presentations and documentary films addressed to a wide variety of neophytes, *i.e.*, collectors, prospectors, students, teacher, researchers and people interested in astro-tourism. In particular, since 2016, the Museum has been committed in organizing a number of workshops that raised the awareness of the audience on how to duly identify and collect meteorites. Noteworthy, this is the first museum exclusively dedicated to meteorites in Morocco, Africa and Arab countries.

The national meteorite heritage represents not only a scientific and educational tool, but also a significant source of income for thousands of inhabitants of the country, who, over time, have developed a factual knowledge in meteorite research and collection. Currently, meteorite hunting has become part of popular culture, and it is common to encounter locals holding magnets, or even metal detectors, in their search for meteorites. Thus, many people have become experts with an essentially economic interest. Finding a meteorite and selling it to foreign buyers is the dream of any prospector. Although the Ibn Zohr University Museum of Meteorites has done an excellent job in ensuring that at least part of this heritage remains in Morocco, nowadays, most Moroccan meteorites are unfortunately kept in collections outside the country.

Fig. (1). Meteorites (**1**) and petroglyphs (**2, 3**) displayed in the University Museum of Meteorites, and Museum visitors (**4**). © University Museum of Meteorites.

METEORITES IN ANCIENT TIMES IN MOROCCO

Ancient Stone Carvings Confirm that Meteorites Struck Morocco

Regardless of the cultures involved, the observation of the sky and astronomical bodies has been of worldwide interest since prehistoric times. Petroglyphs have been found around the world and have been interpreted by researchers as signs of the Sun (Coimbra, 2009), the Moon (Olivera & Silva, 2010) and supernovae (Iqbal *et al.*, 2009). However, very few of them have been interpreted as bolides (Coimbra, 2007), comets (Coimbra, 2010) and meteors or meteorites (Iqbal *et al.*, 2010, Figueiredo *et al.*, 2017). These events have been interpreted by early societies as bad or good manifestations of the gods and, therefore, carved on rocky surfaces to be admired by future generations (Sagan & Druyan, 1986). In particular, Bailey (1995) argues that phenomena related to meteors and comets have played an important role in the beliefs and social habits of most civilizations.

Meteors and meteorites have fascinated humans since they were first spotted in the night sky, but ancient cultures often explained these phenomena based on myths and legends. The study of meteoritic phenomena described in the indigenous oral traditions has been a topic of research interest for several years. In particular, the research in the disciplines of "astromythology" and "ethnoastronomy" involves the investigation of oral traditions in the descriptions of past astronomic events and may provide insights both into the culture that observed and recorded them and into the event itself that may contribute to the understanding of meteoritic phenomena. Meteors and meteorites have been observed by humanity for thousands of years, but only in the past few centuries, they have been studied as astronomical phenomena. Before modern times, great comets caused fear, being considered a bad anticipation of disasters and turmoils, *e.g.*, announcing the death of a ruler (Köehler, 2002).

In Morocco, the astronomical representations engraved on rocks confirm the great attention given by ancient people to the observation of meteors and meteorites. In particular, three pebbles engraved with figures of astronomical bodies such as meteors, anthropomorphic, zoomorphic and inscriptions Tifinagh, were recently discovered by Moroccan researchers and International Meteor Organization members, including the author and his collaborators in the Tiwrare rugged rural area, at heights ranging from 800 to 1500 m (30°59'49.7" N, 9°32'10.9" W), municipality of Ida Oukazou, Essaouira province, approximately 100-km north of Agadir (Fig. **2**). The petroglyphs were located exactly 10 km south of the recently constructed P2234 Tamanar-Adrdor road that favored to reach this location. These petroglyphs, which are currently preserved at the University Museum of Meteorites of Agadir, suggest that ancient Moroccans observed meteorite falls (Ibhi *et al.*, 2018 and 2019). These engravings were compared to other petroglyphs discovered previously in the region, and testimonies gathered from the eyewitnesses of the Tissint meteorite fall in 2011 allowed us to conclude that these petroglyphs date back to ancient time, although no information could be achieved on their age. However, the researchers concluded that the three petroglyphs would represent a meteor followed by the impact of a great meteorite that frightened the inhabitants and that the artist has certainly experienced this astronomical event spectacular enough to be recorded on the rock. In particular, these petroglyphs provide a new perspective on Amazigh archaeoastronomy in Morocco and contribute to the understanding of the region's ancient history.

The artist used three sandstone and quartzitic sandstone pebbles named Ida 1, Ida 2, and Ida 3. The mesoscopic features of which were examined by a binocular magnifying glass equipped with an embedded digital camera to recreate the artist engraving technique, the direction of tool movement, *etc*. The samples Ida 1 and Ida 2 consist of two subcircular-shaped stones of melanocrate cryptocrystalline quartzitic sandstone that measure, respectively, 20 and 18 cm in length, 17 and 15 cm in breadth, and 5 cm in thickness (Fig. **3**). Both samples exhibit a dandruffy surface calcification that completely covers the etched images and consist of thin plates of more or less continuous white or beige color (Fig. **3**), that indicate these objects are ancient.

After cleaning with a soft brush and vinegar, the lone etched side of Ida 1 shows a stunning depiction of a man and woman who appear to be scared by the fall of a meteor (Fig. **4**). Although Ida 2 has not yet been cleared of its clay and sand gangue and still contains secondary precipitates of carbonate dandruff, a scene with a sprinting figure and a giant falling ball can be recognized (Fig. **5**).

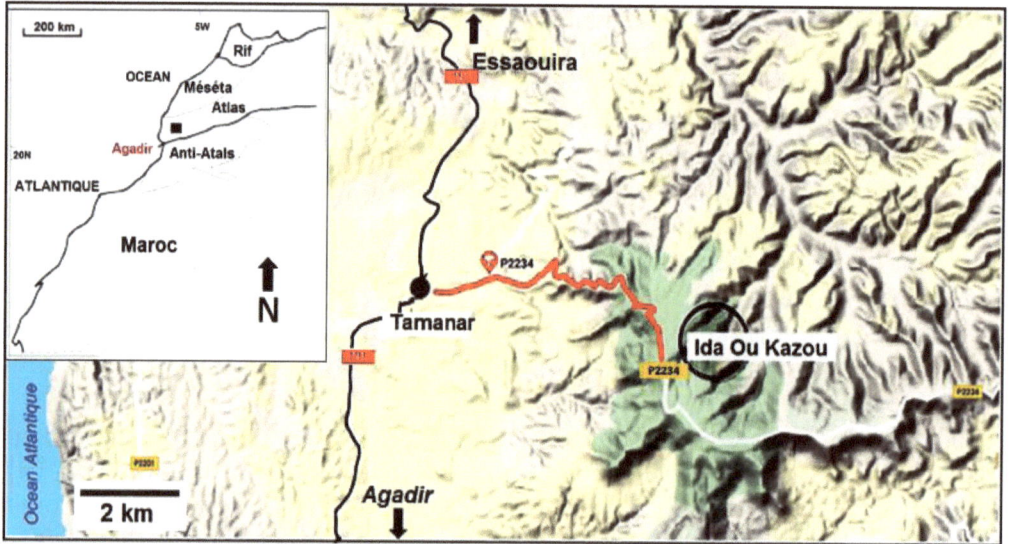

Fig. (2). Geographical location of Ida Oukazou, Essaouira province, Morocco.

Fig. (3). Incised lines and secondary calcification on Ida 2.

Ida 3 is a flat and somewhat square leucocrate sandstone pebble 35-cm long, 27-cm wide and 12-cm thick. The mesoscopic study suggests that the same artist engraved all the figures on this stone, which represent a scenario that shows an anthropomorphic character, two cattle of various sizes, a meteor and the sun with concentric circles in the middle. The artist has also incised two lines of writings in Tifinagh characters in the left area on the sides of the ideogram (Fig. **6**). The Tifinagh writings are ancient and difficult to be dated accurately.

Fig. (4). The petroglyph Ida 1, both before and after cleaning.

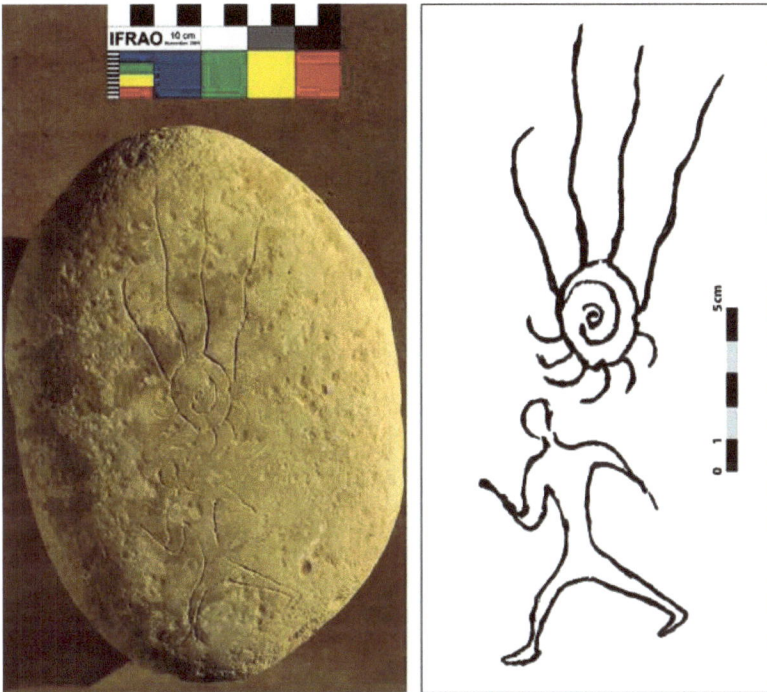

Fig. (5). The petroglyph Ida 2.

Numerous astronomical figures of various chronology have been painted or carved on rocks all over the world, which suggests that one of the primary priorities of prehistoric humans was to examine the sky. Actually, the depiction of celestial entities like the sun, moon and meteor phenomena, may provide an understanding of the thinking processes of these ancient cultures and allows us to interpret how they conceptualized these astronomical facts. In particular, Ida 1 and Ida 2 stone carvings feature concentric circles with hollowed-out centers, are joined to a series of three wavy lines that extend backwards (four for Ida 2), and undoubtedly depict one round object flying in the sky and leaving behind a trail of fire, *i.e.*, a meteor (Fig. 7). The typology of these paintings is very similar to those of the engraving of Toca do Cosmos (Bahia, Brazil) (Coimbra, 2009) and that of the rock painting of the Fouriesbourg district (South Africa) (Woodhouse 1986) (Fig. **8**). The lines engraved on the petroglyphs show long, wavy tails that provide a bright and very dynamic aspect of a flying object. Eyewitnesses of the Tissint meteorite fall occurred in 2011 in the Tata region of Morocco reported that the fireball appeared in the sky with a continuous trail of smoke and dust (Ibhi *et al.*, 2013a). Therefore, the wavy lines engraved on the petroglyphs can be interpreted as the smoke left behind by the meteor.

Fig. (6). The petroplyph Ida 3 before and after cleaning.

Astronomical interpretations suggest that these carvings represent a meteor, and depict the impact of a big meteorite that alarmed the locals. The artist most likely witnessed this astronomical event that he found so spectacular to record it on the rock. Even though the three rock arts depict the same scenario, mesoscopic studies reveal that they are technically and aesthetically distinct from one another (Fig. **7**). There is no reliable evidence that the three comparable pieces were created by the same person at the same period, and probably other similar objects would be present in the region.

A fundamental issue is dating petroglyphs. The archaeological finds we are discussing here have never been studied, thus currently, there is no way to date them in the absence of further research that should include the themes represented, the style and technique of the engraving, the patina, and the history of the Tifinagh inscription. However, chemical dating based on the quantity of certain chemical elements present in the calcitization coatings and the micromorphological analyses of these dandruff thin slides may contribute to determine the age in which these petroglyphs were carved.

Fig. (7). Ida 1: a hollow-center fireball and two human figures (perhaps a female **(A)**, a male **(B)** and the meteor **(D)**. Ida 2: a human and a meteor figure **(C) (E)**. Meteor with four wavy streaks in Ida 3 **(F)**.

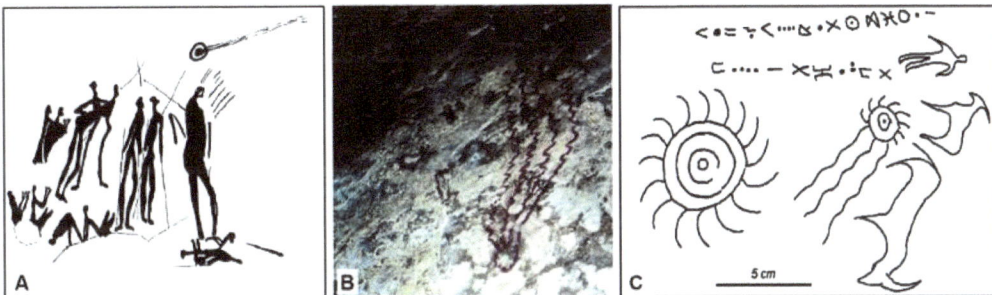

Fig. (8). (A) - Meteor of Toca do Cosmos (Bahia, Brazil) (Coimbra, 2009). **(B)** - Fouriesbourg Meteor (South Africa) (Woodhouse, 1986). **(C)** - Petroglyph Ida 3.

All petroglyphs in arid regions of the world are covered with layers of secondary deposits, *i.e.*, the deposition of carbonates that occurs continuously on an archaeological time scale of hundreds of years (Hassiba *et al.*, 2012). Numerous attempts have been made to date a rock artifact by examining the secondary deposit layers by radiocarbon analysis (Bednarik 2008, Rowe& Steelman 2003). In particular, the carbon included in the carbonate may be utilized to determine the age of the associated rock art.

Prior to the carving of a petroglyph, a carbonate crust would form on the stone exterior as a consequence of natural deposition, as shown in the schematic illustration of the carbonate deposition process in Fig. (**9**). Apparently, the carbonate coating solidified on the quarzitic sandstone substrate of the petroglyph Ida 2 (Fig. **9A**). SEM examination conducted at the Ibn Zohr University revealed that these carbonates are combined in large amounts with dolomite (double carbonate of calcium and magnesium).

The carbonate crust would be removed when the artist worked on the rock (Fig. **9C**). Then, a fresh layer of carbonate crust was naturally deposited on the exposed surfaces. Therefore, the freshly formed crust on the carved surface would indicate the engraving period, as carbonate deposition would have started shortly after the petroglyph was carved (Fig. **9D**).

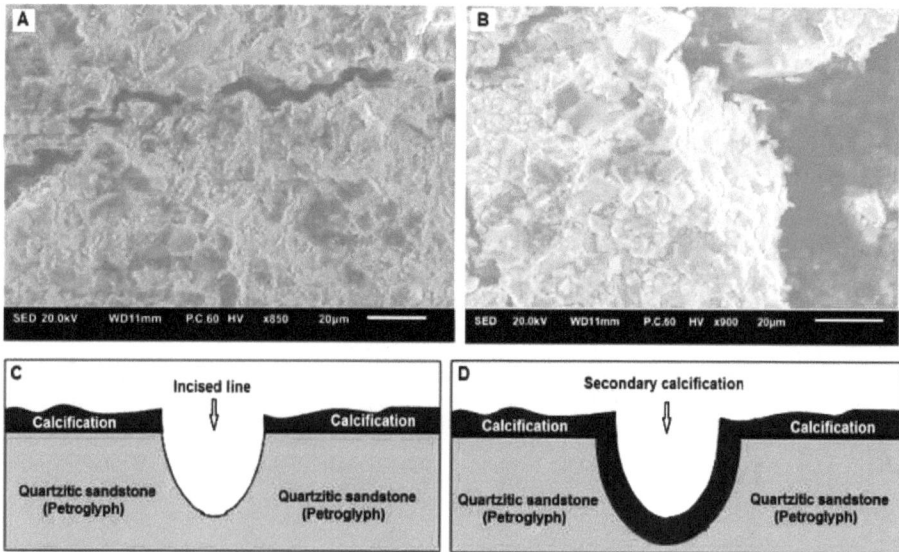

Fig. (9). The Ida 2 dolomite and carbonate film (**A**). The petroglyph Ida 2 substrate is quarzitic sandstone (**B**). Diagrammatic representation of the carbonate deposit process on Ida 2 petroglyph (**C**). The deposit accumulates over time on the areas excised by the artist (**D**), which are used for dating.

Future research is expected to validate or disprove the theory of a meteorite impact, on the basis of the existence of meteorite fragments and impactites (terrestrial rocks affected by the impact) and the presence of metamorphic shock effects in the area where the petrogliphs were found. Further geological, magnetic and gravity prospecting studies are planned in order to verify this theory, which could be applied not only to Moroccan archaeoastronomy, but also to discover possible impact craters in the area.

In conclusion, this preliminary study on the petroglyphs collected in Morocco is considered particularly useful for promoting a discussion among scientists interested in Moroccan archaeoastronomy and contributing to the advancement of scientific studies in this area.

The Meteoritic Origin of Morocco Iron Dagger Blades

A long unresolved and controversial debate exists on the meteoritic (extraterrestrial) or smelted (terrestrial) origin of Bronze Age iron artifacts (Jambon, 2017). Although the Iron Age began in Anatolia and the Caucasus around 1200 BCE, nearly 2000 years earlier, various cultures were already using iron objects. These items were extremely rare and always considered very precious, although iron ores abound on the terrestrial surface.

Only a few detailed scientific researches have reported convincingly the identification of meteoritic iron in ancient artifacts. Among these, two funerary iron bracelets and an axe excavated in two different Polish archaeological sites (Kotowiecki, 2004) and an ancient "iron man" Buddhist sculpture carved from a fragment of the Chinga meteorite (Buchner *et al.*, 2012). According to Comelli *et al.* (2016), the iron dagger blade found in the sarcophagus of King Tutankhamen of Egyptian Civilization (14th century BCE) originates from a meteorite. The chemical composition (Fe plus 10.8 wt% Ni and 0.58 wt% Co) of this blade was properly identified using portable X-ray fluorescence (pXRF), which confirmed its meteoritic source, suggesting that ancient Egyptians gave high importance to meteoritic iron for the creation of valuable objects. At the time of the stabilization of the Earth, almost all Ni drifted towards the core of molten iron. As a result, the rocks on the Earth's surface feature a low concentration of Ni, whereas meteorites resulting from the bursting of small celestial bodies, in particular M-type asteroids, are relatively rich in Ni and Co, which makes it easy to recognize the source of iron artifacts.

Using a pXRF spectrometer, Jambon (2017) revealed that a small number of Bronze Age iron artifacts in museum collections were constructed with meteoric iron, in particular, the beads of Gerzeh (Egypt, -3200 BCE), a knife from AlacaHöyük (Turkey, -2500 BCE), a pendant from Umm El-Marra (Syria, -2300 BCE), an artifact from Ugarit (Syria, -1400 BCE) and others from the Shang dynasty civilization (China, -1400 BCE). A similar study conducted a year later by Chen *et al.* (2018) showed that two Bronze Age daggers collected from central China were made from extraterrestrial iron.

Several iron meteoritic fields have been discovered (Ibhi *et al.*, 2013; Nachit *et al.*, 2013 and Moggi-Cecchi *et al.*, 2015) in the North and North-West Morocco, mainly in the regions of Agoudal (Imilchil) and Maatarka (Taza). Using metal detectors, thousands of fragments ranging from 2 to 1000 g, with the biggest piece weighing 200 kg, were found in the territory of Agoudal by Moroccan and foreign meteorite hunters. The majority of Agoudal meteorite fragments were collected from a 7 x 2.5 km littered field oriented North-South (Ibhi *et al.*, 2013). In the Oglat Sidi Ali region, more than 1500 kg, of which a 190 kg piece, were found in a NE-SW oriented scatter field 20-km long and 5-km wide (Moggi-Cecchi, *et al.*, 2015). Although a wealth of meteoritic iron was found in Morocco, no object made with this extraterrestrial element has been found until four artifacts, out of many dozen of artifacts that a French antique dealer collected in Morocco near the town of Imilchil and Maatarka in 2020 during an expedition for meteorite and antique items recovery, tested positive for Ni using the dimethylglyoxime (DMG) test. However, as many industrial iron items include enough Ni to get a positive result on the DMG test, this test may yield "false results", which make it insufficient for a doubtless determination of their meteoritic origin.

In a recent study (Ibhi *et al.*, 2022), the four iron dagger blades mentioned above (Fig. **10**) were analyzed by pXRF and scanning electron microscopy-energy dispersive X-ray spectroscopy (SEM-EDS) using a JEOL JSM IT1000 at the Scientific Research Center of the Faculty of Sciences of the Ibn Zohr University. Artifact A (dimensions, 30 cm x 2.9 cm x 0.1 cm; weight, 112.9 g) appeared not severely weathered and showed relatively recent manufacture, whereas artifacts B (dimensions, 23.3 cm x 2.5 cm x 0.3 cm; weight, 98.5 g), C (dimensions, 14 cm x 1.5 cm x 0.2 cm; weight, 27.6 g) and D (dimensions, 8.5 cm x 1-1.3 cm x 0.2 cm; weight, 15.1 g) appeared very altered (Fig. **10**) by weathering effects due to their environment and age.

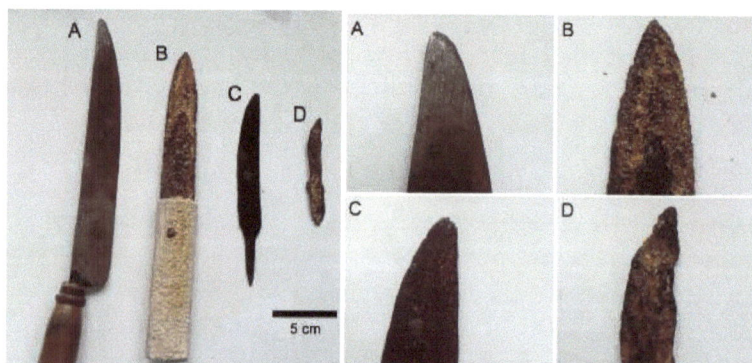

Fig. (10). The four-blade daggers analyzed. Left: full-size images. Right: detailed images exhibiting strongly oxidized rusty and weathered surfaces.

In particular, knife D, whose handle and blade are made of metal, does not contain Ni and Co; knife B clean surface revealed that it is made up mostly of hydrated Fe oxides with a median Ni concentration of 0.95 wt%, does not contains Co and the concentrations of Ca, Mn and Cu are below 1 wt% and likely to originate from the contaminating soil, whereas knife C featured 7.2 wt% of Ni and 1.1 wt% of Co, and intergrowth lamellas (from 20 to 100 µm) of kamacite and taenite (Ni-rich, 34.5 wt%) were visible by SEM.

Furthermore, the pXRF measurements carried on the surface of the dagger blade C showed that Fe and Ni are its main bulk constituents, whereas the other three blades contained only < 0.95 wt% Ni, so they were not of meteoritic origin.

The possible genetic relationship of blade C with iron meteorites was evaluated by comparison with two meteorite samples collected in Agoudal and Oglat Sidi Ali meteorite fields located respectively 50 and 100 km from the place where the daggers were recovered. The SEM-EDS analysis of Agoudal meteorite showed that it mostly included schreibersite and kamacite with average Ni and Co concentrations of 5.5 wt% and 0.4 wt%, respectively (Ruzicka *et al.*, 2015). The ICP-MS analysis of the Oglat Sidi Ali meteorite showed a Ni content of 14.7 wt%, very similar to that of blade C, and a Co content of 1.1 wt% (Bouvier *et al.*, 2017). The SEM analyses of this meteorite also featured fine lamellae consisting of kamacite (mean Ni content of 67 ± 0.1 mg/g) and taenite (mean Ni content of 147 ± 0.2 mg/g) (Moggi Cecchi *et al.*, 2015). Thus, the OglatSidi Ali meteorite mineralogy and plessitic texture are similar to those of blade C, also because the plates of kamacite are segregated in plessite fields, whereas the Agoudal meteorite showed no mineralogical or textural relationship with this dagger (Ibhi *et al.*, 2022).

In conclusion, the percentages of the two key main elements, Ni and Co, in the three dagger blades, A, B, and D, were within the range of levels typical of terrestrial iron minerals, but those in specimen C, together with its chemical, microstructural composition, were typical of extraterrestrial iron, most likely a fragment of the Oglat Sidi Ali meteorite.

The discovery of this iron artifact in the Moroccan High Atlas suggests that extraterrestrial iron was used as the metal source during the country historic iron mining. This study is the first scientific testimony concerning the use of meteoric iron as a source of metal in Moroccan metallurgical industries. The University Museum of Meteorites and Moroccan collectioners should work together to establish techniques of investigation for man-made meteoritic artifacts (Ibhi *et al.*, 2022). Kotowiecki (2004) argues that, despite being altered, meteoritic artifacts can be recognized as objects of extraterrestrial origin.

MOROCCAN METEORITE FALLS

Historically, the strongest evidence for the extraterrestrial origin of meteorites is observed falls, however, according to Khiri *et al.* (2015, 2017) and Ouknine *et al.* (2018), data based on observed falls and meteorite discoveries are difficult to interpret and typically rely on ad-hoc assumptions about the distribution of human populations and their various levels of education.

This contribution lists in chronological order each well-documented meteorite fall in Morocco, which has been observed by eyewitnesses as it passed through the sky or touched the soil. At the end of December 2021, 22 official records (21 well documented) for observed meteorite falls have been counted in Morocco, a list of which is provided in Table 1, which includes meteorite names, their fall dates, locations, and types, together with the name of the corresponding meteors after the place they fell. The majority of the data in Table 1 was extracted from the Bulletin and the database of the Meteoritical Society.

Table 1. Moroccan meteorite falls in chronological order.

Year	Name	Latitude	Longitude	Date	Type	Weight
1932	DouarMghila	32° 20'N	6° 18'W	Aug 20	LL6	1161 g
1986	Oued el Hadjar	30°10'48"N	6°34'38"W	Mar	LL6	1216 g
1990	Itqiy	26°35'27"N	12°57'8"W	March	EH7	4.72 kg
1998	Zag	27°20'N	9°20'W	Aug 4 or 5	H3-6	175 kg
2002	Benssour	30°N	7°W	Feb 11 (4:30 GMT)	LL6	~45 kg
2003	OumDreyga	24°18'N	13°6'W	Oct 16	H3-5	17 kg
2004	Benguerir	32°13'52.9"N	8°8'56.7"W	Nov22 (11:45 GMT)	LL6	25-30 kg
2008	Tamdakht	31°9'48"N	7°0'54"W	Dec 20 (22:37 GMT)	H5	~100 kg
2011	Tissint	29°28.917'N	7°36.674'W	Jul 18 (02:10 GMT)	Sher	20 kg
2012	Aoussred	22°18'34"N	14°06'293W	May 20 (22:45 GMT)	CH	270 g
2014	Tirhert	28.935°N	8.905°W	July 9 (22:30 GMT)	Eu-un	8-10 kg
2014	Tinajdad	31°36'33.2"N	5°11'38.7"W	Sept 9 (7:30 GMT)	H5	1860 g
2015	Sidi Ali OuAzza	29°47'2.9"N	7°23'21.8"W	Jul 28 (18:30 GMT)	L4	1500 g
2016	OudiyatSbaa	25.546°N	12.418°W	Nov 18 (16:40 GMT)	EH5	23,85 kg
2018	Matarka	33°15'N	02°45'W	Jan 5 (21:30 GMT)	L6	538 g
2018	Gueltat Zemmour	25°05'32.0"N	12°37'23.9"W	Aug 21 (14:20' GMT)	L4	8 kg
2018	Ksar El Goraane	32°11'13.2"N	03°42'04.6"W	Oct 28 (22:30 GMT)	H5	4 kg
2019	Wad Lahteyba	27°22'23.15"N	8°58'51.7"W	Jun 27 (17:00 GMT)	H5	20 kg
2019	Al Farciya	27°01'27.59"N	9°44'39.59"W	Aug 20 (01:15 GMT)	L6	1300 g

(Table 1) cont.....

Year	Name	Latitude	Longitude	Date	Type	Weight
2020	Tarda	31° 49' 35"N	4° 40' 46"W	Aug 25 (14:30 GMT)	C2-ung	4 kg
2021	Tiglit	28°23.533'N,	10°22.632'W	Dec 10 (19:45 GMT)	Aubrite	~2 kg

The first one, the Douar Mghila (LL6), was observed in 1932, and no other fall was reported during the next 54 years. The Oued El Hadjar (LL6) fell in 1986, followed by Itqiy (EH7- anomalous) in 1990 and Zag (H3-6) in 1998. Since 2002, the news falls submitted to the Nomenclature Commission are Bensour (LL6, 2002), OumDreyga (H3-5, 2003), Benguerir (LL6, 2004), Tamdakht (H5, 2008), Tissint (Martian shergottite, 2011), Tirhert (Eucrite unbrecciated, 2014), Tinajdad (H5, 2014), Sidi Ali OuAzza, (L4, 2015), OudiyatSbaa (EH5, 2016), Matarka (L6, 2018), GueltatZemmour (L4, 2018), Ksar El Goraane (H5, 2018), Wad Lahteyba (H5, 2019), Al Farciya (L6, 2019), Tarda (C2-ung, 2020) and Tiglit (Aubrite, 2021). The geographical distribution of Moroccan falls is shown in Fig. (**11**).

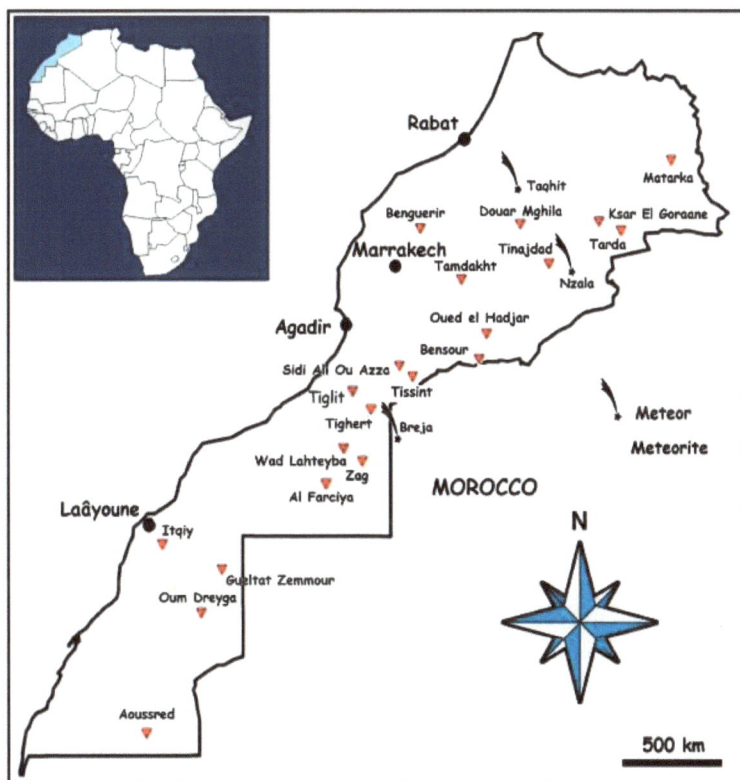

Fig. (11). Distribution of meteorite falls in Morocco.

Exceptional Meteorite Falls

– *DouarMghila* (1932, ordinary chondrite LL6, 32°20'N, 6°18'W) (Fig. **12a**). A kind of rocket crossing the sky from east to west was observed, accompanied by a detonation similar to a gun which was heard at a distance of 10 km from the area where forty stones fell, but only 2 were recovered with masses of 846 g and 315 g. Captain Larcher, who saw this occurrence, claimed that the event occurred 4 km east of BeniMellal (Lacroix, 1933). The mineralogical composition of major silicates was olivine, orthopyroxene, plagioclase, diopside and accessory chromite. A Pb, Th/He gas retention age of 3.13 Gyears and a cosmic ray exposure (CRE) age of 17.0 Myears have been reported (Grady, 2000). Most of the mass was gained by the "Musée National d'Histoire Naturelle" (Paris), with smaller specimens by the Natural History Museum (London) and the Scientific Institute at the Mohammed V University of Rabat.

– *Zag* (1998, ordinary chondrite H3-6, 27°20'N, 9°20'W) (Fig. **12b**). The meteorite fall was witnessed on a mountain in the vicinity of Zag, Morocco. About 175 kg have been sold by local people to dealers and collectors under the names Zag, Sagd, and Tan-Tan (Grossman & Zipfel, 2001). The Zag meteorite is the second meteorite after the Monahans meteorite featuring the presence of halite crystals containing microscopic inclusions of water. Both finds have raised hopes of learning more about the possibility that life might have evolved elsewhere in the solar system (Singmaster, 2000). The meteorite is a regolith breccia containing olivine (Fa1.6–30), and pyroxene (Fs3.3–26.6); the shock stage is S3 and the weathering grade W0-1.

Fig. (12). Exceptional meteorite falls in Morocco. **(a)** DouarMghila (©MNHNP), **(b)** Zag (©Meteorite.fr), **(c)** Tissint (©Ibhi A.), **(d)** Tighert (©Ibhi A.), **(e)** Tarda (©M. Oulkouche), **(f)** Tiglit (©Ibhi A.).

– *Tissint* (2011, Shergottite 29°28.917'N, 7°36.674'W) (Fig. **12c**). The meteor entered the atmosphere in the south-east skies of Tata, Morocco. Its contact with the atmosphere produced brilliant light flashes accompanied by detonations. A large number of fragments survived the fire ball phenomena (Ibhi, 2012a, 2012b). Dozens of fragments with masses ranging from 0.2 to 1.280 g were collected, totaling roughly 12–15 kg. The meteorites collected are variably coated by a shining black fusion crust, and the exposed interior appears pale green-grey in color, with mm-sized, pale-yellow olivine phenocrysts with sparse vesicular pockets and thin veins of black glass (Ibhi *et al.*, 2013a). The meteorite was ejected from the surface of Mars between 700,000 and 1.1 million years ago and appeared to be derived from a deep mantle region different from any of the other known Martian shergottite meteorites (Schulz *et al.*, 2020). The overall composition of the meteorite corresponds to that of an aluminum-poor iron basaltic rock that likely originated from magmatic activity on Mars's surface. The basalt then underwent weathering by fluids that deposited minerals enriched in incompatible elements in fissures and cracks. A later impact on the surface of Mars melted the leached material forming black glassy veins.

– *Tighert* (2014, unbrecciated gabbroic eucrite, 28.935°N, 8.905°W) (Fig. **12d**). This meteorite fell on July 9, 2014, at about 10:30 pm. Across its atmospheric course, the bolide underwent many fragmentation events as it moved from north to south-east. Eyewitnesses reported seeing the bolide and hearing explosions a short while afterwards at a number of locations in the Tirhert, Assa, Tata, DouarImougadir, Taghjijt, Foum El Hisn, and Guelmim-Es-Semara area. Soon after the ball of fire incident occurred, the local authorities conducted a field assessment to solve any potential security issue. The thorough mineralogical and petrological analyses of the meteorite have shown that it is similar to a eucrite "magmatic" meteorite from the asteroid belt, namely Vesta-4 (Ibhi, 2014a). Eyewitnesses provided information on the locations of several fragments that formed a scattered strewnfield about 63 km in length in the direction NW to SE. The biggest mass found was gathered not far from Tighert, and weighed about 1300 g. With an estimated total mass of 8 to 10 kg, the recovered fragments ranged in weight from 1 to 1300 g. Most of them show an extremely glossy black fusion crust. Plagioclase and pyroxene grains, as well as a few opaque grains, appear on the fractured surfaces. The magnetic susceptibility is log $\chi = 2.53$, and the general texture is equigranular (Ruzicka *et al.*, 2017).

– *Tarda* (2020, Carbonaceous chondrite (C2, ungrouped, 31°49'35"N, 4°40'46"W) (Fig. **12e**). On Tuesday, August 25, 2020, at about 02:30 pm, a fireball was widely witnessed by people in Alnif, Zagora, Tazarine and Rich, in southern Morocco. Hundreds of people began searching on the same day, and the first piece was found the following day close to the village of Tarda. The fall location is crossed by the national road linking Ouarzazate to Errachidia, and because of the easy access, thousands of people soon traveled to the area. Several eyewitnesses reported "a bright yellow, barrel-sized fireball with green edges", which released little pieces around its edges. Thousands of small fragments ranging from a few mg to 99 g were collected. Freshly picked-up stones were reported to have a charcoal-like odor. The interiors of the stones are dull black with dispersed white or light-colored grains or clasts. Optical observations revealed a matrix-rich breccia containing small chondrules (granular, BO) and chondrule fragments, very fine-grained AOA, and grains of forsterite. No CAIs were identified. Electron microprobe surveys of thin sections and powder X-ray diffraction studies have shown that the matrix is dominated by phyllosilicates with low amounts of magnetite, pyrrhotite, pentlandite, troilite, carbonates and olivine. The isotopic values of the meteorite do not correspond to those of any established carbonaceous chondrite group, hence the ungrouped designation. Magnetic susceptibility is log χ = 4.99 (Meteoritical Bulletin, no. 109).

– *Tiglit* (2021, Aubrite, 28°23.533'N, 10°22.632'W) (Fig. **12f**). On the evening of December 10, 2021, at about 07:45 pm, a bright bolide was observed by thousands of eyewitnesses in an area about 140 km south-east of Guelmim town (South-East Morocco). A terminal fragmentation and sound phenomena were perceived near the end point of the trajectory. The bolide traveled from northwest to southeast and experienced several fragmentations along its trajectory. The first fragment was discovered the next day. About a hundred people came to the region of Tiglit (Tata, Morocco) and thousands of fragments were collected by nomads, traders and hunters (Ibhi, 2022). The meteorite features a monomict breccia structure typical of aubrites, which consists of coarse-grained enstatite fragments up to 5 mm in size with a fine-grained matrix. Enstatite (Opx) with diopside (Cpx), olivine, glass and opaque minerals compose the matrix, whose phases are typically tiny (less than 100 µm in size). The enstatite is almost homogeneous in composition (very rich in magnesium and poor in iron), olivine is pure forsterite (very poor in iron) and clinopyroxene is a diopside, with minor metals, high in Ni, associated with troilite. Based on these data, this meteorite should be classified as an achondrite meteorite of the "Aubrite" type, and not as a lunar feldspathic regolith breccia as proposed by merchants (Ibhi, 2022).

Ordinary Chondritic Meteorite Falls

– *Oued el Hadjar* (1986, Ordinary chondrite LL6) (Fig. **13a**). The meteorite fell in March 1986 in the Guelmim-EsSemara region. Witnesses heard a whistling sound, and a single piece fell less than 30 m far from their tent. The meteorite was found days later broken into fragments, and one of them was used as an offer on a traditional wedding altar. Fragments of the original specimen were acquired later by a meteorite dealer, but some of the meteorites was lost and never recovered. The mineralogical features and classification are: olivine (Fa30.0), pyroxene (Fs22.0), shock stage S2, and weathering grade W0-1 (Grossman and Zipfel, 2001).

Fig. (13). Ordinary meteorite falls in Morocco. **(a)** Oued el Hadjar (©Matteo Chinellato); **(b)** Itqiy (©Piotr gural); **(c)** Bensour (©Matteo Chinellato); **(d)** Oum Dreyga (©Bartolommei-giulio); **(e)** Benguerir (©Jay Piatek); **(f)** Tamdakht (©DCOM); **(g)** Sidi Ali Ou Azza (©Woreczko Jan & Wadi); **(h)** OudiyatSbaa (©A. Irving); **(i)** Matarka (©Beat Booz); **(j)** Ksar El Goraane (©Hanno Strufe); **(k)** GueltatZemmour (©Wang Ziyao); **(l)** Wad Lahteyba (©Meteorite Belgum).

− *Itqiy* (1990, Ungrouped enstatite-rich meteorite) (Fig. **13b**). After an explosion and the appearance of light, a nomad found one fragment weighing 410 g close to Itqiy. While looking for meteorites in the same location in July 2000, the French collectors Marc Luc and Jim Labenne discovered a larger fragment weighing 4310 g. The meteorite shows an equigranular texture with numerous triple junctions. Other features are: enstatite 78%vol (Fs0.2, Wo3.0), kamacite 22%vol, with Fe = 90.4%wt, Ni = 5.77%wt, troilite nearly absent, no relic chondrules present; shock stage S5, weathering grade W1–2 (Grossman & Zipfel, 2001).

− *Bensour* (2002, Ordinary chondrite LL6) (Fig. **13c**). In the border region between Morocco and Algeria, several nomads eyewitnesses reported a huge fall of several fragments, however these accounts lacked specificity. The total weight is likely more than 45 kg, the biggest being a 9.2 kg fragment that crashed into three large pieces on impact. The meteorite consists of a very fine-grained monomict breccia composed of olivine (Fa31.6), orthopyroxene (Wo3.5, Fs24.3), clinopyroxene, sodic plagioclase with subordinate Ni-rich Fe-Ni metals, troilite and chromite featuring sparse chondrules (Grossman & Zipfel, 2001).

– *OumDreyga* (2003, Ordinary chondrite H3-5) (Fig. **13d**). On 16 October 2003, in the GourLafkah Mountains, south of Zbayra and about 21 km from OumDreyga, Moroccan soldiers placed in Western Sahara witnessed a meteorite impact. The meteorite fell near a 670-km-long sand wall built in 1985 which was protected by antipersonnel mines and guarded by soldiers. Soon after the fall, some pieces were found with a total weight of 17 kg. Some fragments have been offered for sale under the names of Amgala and Gor Lefcah. Classification and mineralogy: very fresh, with a black fusion crust, H3-5 breccia, S4, W0 (Russell *et al.*, 2005).

– *Benguerir* (2004, Ordinary chondrite LL6) (Fig. **13e**). On November 22, 2004, at about 11:45 GMT, a meteorite shower was witnessed by local people to fall near Benguerir (about 50 km north of Marrakesh). The estimated total recovered mass is about 25-30 kg. Three fragments fell at Douar Lfokra (32°13'52.9" N, 8°8'56.7" W), Ahl Foum Sakhra Lourania (32°15'31.2" N, 8°10'51.9"W) and Douar Tnaja (32°15'43.1"N, 8°9'1.3"W). The meteorite shows a dull black fusion crust and regmaglypts. Mineralogy and classification: LL6, Fa29±1, Fs25±1, S3, W0 (Russell *et al.*, 2005).

– *Tamdakht* (2008, ordinary chondrite H5) (Fig. **13f**). On December 20, 2008, witnesses reported seeing a meteor with a west-to-east trajectory over a number of localities (Agadir, Marrakesh, and Ouarzazate). Residents of the high Atlas Mountains between Marrakesh and Ouarzazate heard a bang and felt an aftershock. The first reports on finding pieces of the meteorite were recorded a couple of weeks later. The largest impact pit (1.10 m in diameter and 70 cm in depth) is located near OuedAachir (31°09.8'N, 7°00.9'W), with a recovered piece weighing more than 30 kg and many small fragments totaling a weight of about 100 kg. A strewnfield of at least 25-km long and 2-km wide was observed.. The meteorite features a thick fusion crust, locally more than 1 mm, and abundant chondrules of size between 0.1 and 1.5 mm with visible but not well-delimited outlines. The mineralogy shows dominant olivine (Fa18) and orthopyroxene (En83, Fs16) with plagioclase, Ca-phosphate, chromite, kamacite and taenite. Classification: S3, W0 and log χ = 5.3 (Weisberg *et al.*, 2009).

– *Tinajdad* (2014, Ordinary chondrite (H5), 31°36'33.2" N, 5°11'38.7" W). On Tuesday, September 9, 2014, at about 07:30 pm local time, a fireball was seen and heard by many people in the southern part of Morocco. Many eyewitnesses from Alnif, Tineghir, Tinajdad, Todra Gorges, Azag N-Ouchchan and other villages described a yellow-colored, bright object moving horizontally for about 2 sec and a sound like thunder. A piece weighing 1860 g (broken into 1 big piece and a few small ones) was found the day after the fall about 16 km WNW of Tinajdad city. The fragments were mostly covered by a black fusion crust with a grey interior. The silicate grains are homogeneous in size, with very few chondrules (barred olivine). The metal and sulphide grains are very small. Mineralogy: olivine (Fa19.5), orthopyroxene (En81.6±0.2, Fs16.9), plagioclase (An10-25, Ab72-84), chromite, kamacite, and taenite. Classification: W0, log χ = 5.29 (Ruzicka *et al.*, 2017; Meteoritical Bulletin, No. 103).

− *Sidi Ali OuAzza* (2015, Ordinary chondrite (L4), 29°47'2.9"N, 7°23'21.8"W) (Fig. **13g**). On Tuesday, July 28, 2015, at about 05:30 pm local time, many people from Tissint and its vicinity heard three sonic booms. Two days after the fall, many small pieces from 5 to 109 g were recovered close to Sidi Ali OuAzza and El Kharoua', a few km west of Tissint. The strewnfield is around 4-km long, and the meteorite fall occurred in a north-to-south direction. The meteorite features a patchy fusion crust that is black and brown in color. Numerous discrete chondrules and fine-grained metal/sulfide are visible. The microprobe examination reveals several porphyritic chondrules and a small number of BO chondrules consisting of olivine (Fa25.7), enstatite (Fs2.1±0.2, Wo0.6) and plagioclase (An8.3, Ab86.9, Or4.9) (Bouvier *et al.*, 2017a).

− *Oudiyat Sbaa* (2016, Enstatite chondrite (EH5), 25.546°N, 12.418°W) (Fig. **13h**). On November 18, 2016, at 04:40 pm local time, a bright fireball was witnessed by many observers traveling from north to south over an area extending from Tiznit to Guelmim, Tan Tan and Smara. A search for fallen stones begun, and on November 23, a shepherd found three black fusion-crusted fragments on sandy soil near Oudiyat Sbaa. When first found, the stones emitted a sulfurous odor. Some well-formed chondrules are present within a recrystallized matrix containing abundant metal and sulfides. Besides dominant enstatite, the specimens contain alkali feldspar, kamacite, niningerite, oldhamite, troilite and schreibersite. Mineralogy: enstatite (Fs0.4-0.6, Wo0.1-0), alkali feldspar (Ab88.7-93.1, An1.1-0.9), log χ = 5.74 (Gattacceca *et al.*, 2019).

− *Ksar El Goraane* (2018, Ordinary chondrite (H5), 32°11'13.2"N, 003°42'04.6"W) (Fig. **13j**). On October 28, 2018, at about10.30 pm local time a fireball, first yellow then red, relatively low in the sky and directed from east to west was observed by many people from southeastern and eastern Morocco. First five sonic booms were heard, followed by a whistling sound, and then by a noise comparable to the unloading of a truck of rocks. A day after, in the morning, the first piece weighting 185 g was recovered at about 5 km east of the Ksar El Goraane village. Many small pieces were found east of the village on the Essnam plateau. The strewnfield is long, about 15 km in the direction ENE to WSW. The meteorite was totally covered by black shiny fusion crust. Small pieces show a primary and secondary fusion crust. The texture is fine-grained and homogeneous, and small-sized chondrules are visible. The microprobe examination shows numerous porphyritic chondrules in a recrystallized groundmass with ubiquitous plagioclase grains and abundant FeNi-metal and troilite. Mineralogy: olivine (Fa19.0), low-Ca pyroxene (Fs16.5, Wo1.4), log χ =5.28 to 5.42 (Gattacceca *et al.*, 2020).

– *Matarka* (2018, Ordinary chondrite (L6), 33°15'N, 02°45'W) (Fig. **13i**). On the evening of January 5, 2018, between 09:00 and 10:00 pm local time, a bright green-colored fireball was witnessed by local people in northeastern Morocco. The event was described for several seconds duration, with no sound phenomena. On February 5, one month after the event, a single piece of 538 g was found near Matarka village. The meteorite shows an angular shape with smoothly rounded edges, indicating significant ablation and extended atmospheric flight. The fragment displays a fresh fusion crust, and the matrix appears bright ash colored. Chondrules are present only as relicts, and the mineral grains (except Fe metal) show strong irregular fracturing. The mean size of plagioclase grains is 80-150 μm. Mineralogy and other features: olivine (Fa23.8-24.7), pyroxene (Fs20.1-20.5, Wo1.1-1.7), plagioclase (An9.9-10.9, Or3.9-7.9), shock stage S3, no weathering (W0) (Meteoritical Bulletin, no. 109).

– *Gueltat Zemmour* (2018, Ordinary chondrite (L4), 25°05'32.0"N, 12°37'23.9"W) (Fig. **13k**). On August 21, 2018, at about 02:20 pm local time, many people in the south of Morocco saw a bright fireball. Searching began ten days after the fall, and the first pieces were found in mid-September, a few km west of Gueltat Zemmour. The direction of the fall was NNE-SSW, and the strewnfield is about 12-km long. A few small pieces totaling around 500 g and 3 bigger ones (5448 g, 1374 g and 426 g) were collected. The fragments were not weathered, totally covered by a black fusion crust, and showed very friable fine grains and numerous easily discernible well-defined chondrules (POP, PO, GPO, PP, RP and glassy chondrules) of size up to 3 mm. Mineralogy: microcrystalline mesostasis, olivine (Fa26.0), low-Ca pyroxene (Fs20.7, Wo1.1), very few plagioclases (An8.2, Or4.8), taenite (Ni 50.3) (Gattacceca *et al.*, 2020).

– *Al Farciya* (2019, Ordinary chondrite (L6), 27°01'27.59"N, 9°44'39.59"W). In the night of 19-20 August, 2019, at 01:15 am local time, a fireball was seen by many nomads in the Al Farciya area. In the morning of the same day, many hunters went to the area where the fall was expected, and the first pieces were found in the evening. Collected pieces show a strewnfield of about 8×1 km from north to south, in good agreement with the witnesses. The fragments were mostly covered by a fusion crust, with the internal part featuring brecciation with dark and clear grey zones. Some mm-sized sulfide grains, thin metals and some chondrules are detected, shock veins and a kind of layering are also visible. The microprobe examination shows ubiquitous apatite and chromite and scattered plagioclase grains with sizes up to 100 μm in diameter, FeNi-metal and troilite. Mineralogy and classification: olivine (Fa25.3), low-Ca pyroxene (Fs21.3), S3, W0, log χ = 4.94(Meteoritical Bulletin, no. 109).

– *Wad Lahteyba* (2019, Ordinary chondrite (H5), 27°22'23.153"N, 8°58'51.744"W) (Fig. 13l). A big meteor appeared on Thursday, June 27, 2019, at about 05:00 pm local time in the sky over the Lhmada region of Zag city, exactly over the rural municipality of Lhtiba. According to eyewitnesses, a ball of fire at first orange-blue in color, then turned to red, illuminating the entire region, and splitting into fragments following a massive explosion appeared, leaving behind a trail of white smoke. The fireball was seen by residents of towns and villages more than 150 km from the site of the fall (Ibhi, 2019). Hundreds of hunters and nomads went to the area, and the first pieces were recovered on June 29, the total mass collected was about 20 kg, with the biggest one weighing 4330 g. A very fresh, thin, black fusion crust covers the pieces, with a grey interior showing numerous chondrules. The microprobe examination shows numerous porphyritic chondrules. Mineralogy and classification: olivine (Fa18.6), low-Ca pyroxene (Fs16.4, Wo1.6), S3, W0, log χ = 5.29 (Meteoritical Bulletin, no. 109).

Strewnfields

Only two of the 20 falls described above were subjected to a strewnfield study, *i.e.*, Tamdakht and Tissint (Ibhi, 2013b). According to eyewitness reports and finder GPS data, the approximate dimension of the Tamdaght strewnfield is about 2-km wide and 25-km long (Fig. 14A, red area). The horizontal trajectory, the shallow impact pits in the loose gravel soil and the total destruction of masses, which did not impact rocks but clay surfaces, provide evidence that the meteorites retained at least some of their cosmic velocity until they hit the ground at an altitude of 1500 m (Ibhi 2013a; 2013b).

According to nomads' observations, the Tissint meteorite flight was in a west-north-west to east-south-east direction, and the strewnfield was 15-km long, covering an area of around 60 km^2 (Fig. 14B). More than 50 pieces of the meteorite totaling roughly 20 kg were found in the strewnfield (Ibhi, 2012c).

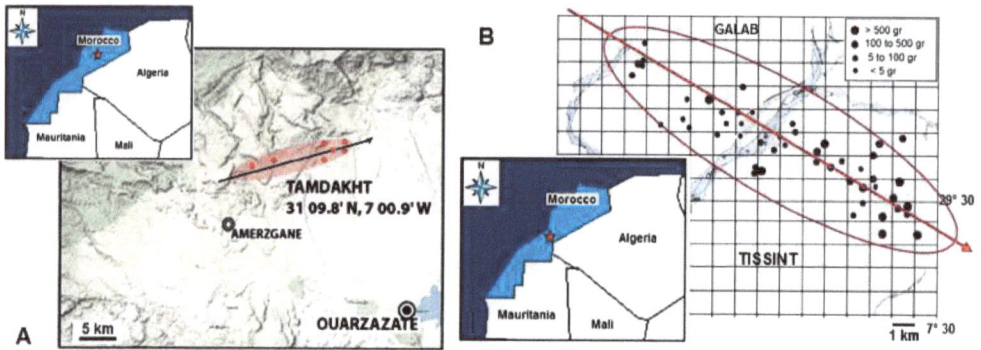

Fig. (14). A: Tamdakht strewnfield (Ibhi, 2014b). B: Tissint strewnfield (Ibhi, 2012b).

Probable Meteors and Falls

At least 6 more meteor detections are reported to have occurred in Morocco, but the exactness of the witnesses' statements is doubtful, and no official meteorite was recovered. These observed meteors and corresponding falls are located in the areas of Nzala (H5, 2009), Breja (LL6, 2010), Izarzar (H5, 2012), Al Mahbes (LL6, 2013) and Kheneg Ljouâd (LL5/6, 2017). However, eyewitnesses have always missed the declaration of the essential elements of these falls, such as the precise location and the exact date of the fall.

1- *Nzala* (2009? Ordinary chondrite). On November 13, 2009, at 09:00 pm GMT, a large number of individuals in Erfoud, Errachidia and Errich saw a meteor and heard three sonic booms. Only two meteorite fragments of about 100 and 700 g were recovered by nomads after weeks of searching (Ibhi, 2014b).

2- *Breja* (2010, ordinary chondrite LL6). On May 1, 2010, at 03:00 GMT in the area of Breja, 35-km north of Tindouf and east of Zag, several people witnessed a very bright bolide accompanied by a sonic boom and a fragmentation event. The meteoritic fragment was found by nomads near the Moroccan border, providing only approximate information on the fall place. The meteorite features S3, W0 and TKW of about 16 kg (Ibhi, 2014b).

3- *Aousred* (2012, Carbonaceous chondrite CH). According to one of the three eyewitnesses, the initial ball was white-orange before becoming red, lit the whole region, and burst into fragments after a significant explosion. The staff at the University Museum of Meteorites were called by a soldier who discovered the first piece (Fig. 15a) (Ait Kadi, 2012; Ibhi, 2014b). The petrographic and mineralogical study of two fragments performed at the Ibn Zohr University Laboratory shows that the meteorite is an H-type carbonaceous chondrite and that the fusion crust surrounding the meteorite was still intact.

4- *Al Mahbes Arraid* (2013, Ordinary chondrite LL6, 27°38.333'N, 9°28.124'W). A fireball fall and three sonic booms were reported in the southern part of Morocco by many eyewitnesses from Assa, Zag and Al Mahbes areas on December 9, 2013, at about 06:30 pm GMT. At the beginning of September 2014, a young nomad girl found two meteorites. To date, the total known mass of the fragments is about 5,500 g, they show a fusion crust, and the interior is brecciated with clear and dark-grey clasts (Fig. **15b**). Neither troilite nor metal grains are clearly visible. The polished mount shows a recrystallized matrix with a few barred chondrules (Gattacceca *et al.*, 2020).

5- *Izarzar* (2015, Ordinary chondrite H5, 30°08.33'N, 07°57.75'W). Residents of the localities of Izarzar and Beni Yacoub reported being awakened by a thunder-like roar followed by a tremor on October 23, 2012, at about 00:30 GMT. Residents from the southern Moroccan cities of Tata, Ighrem, Taghmout and Faddouks also reported to have seen the fall. On October 30, the first piece was found on the Azaghzaf Mountain, about 6 km southwest of Izarzar. The fragments were covered by a dull-gray fusion crust, and showed chondrules sized from 0.1 to 1 mm (Fig. **15c**). Mineralogy and other features: large olivine phenocrysts, microcrysts of diopside and interstitial plagioclase, abundant mm-sized metal patches (kamacite with subsidiary taenite), minor merrillite, chromite and troilite (Bouvier *et al.*, 2017b).

6- *Kheneg Ljouâd* (2017, ordinary chondrite LL5/6, 28°59'03.3''N, 8°24'38.7''W). On the evening of July 12, 2017, at 10:10 pm GMT, a bright bolide was observed by thousands of eyewitnesses in an area 140-km south of Tata town. The bolide traveled from north to south and experienced several fragmentation events along its trajectory. This extraordinary and rare event is the brightest and most comprehensively observed fireball in Morocco's astronomical history and is extremely valuable to the scientific community (Ibhi *et al.*, 2017). The total mass collected is about 10 kg with pieces of sizes from <1 g to 1.2 kg. The fusion crust is black (Fig. **15d**), and the interior is largely whitish grey, with a few thin shock veins and veinlets of troilite (Gattacceca *et al.*, 2017).

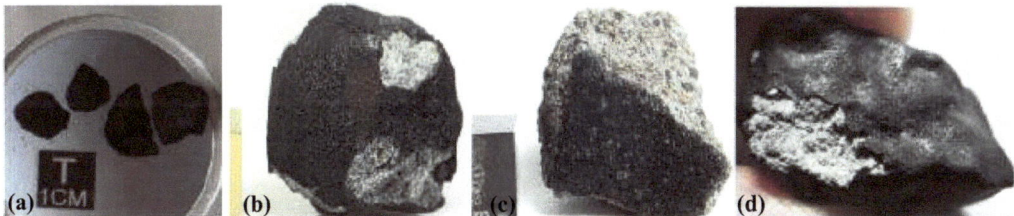

Fig. (15). Probable falls in Morocco. **a.** Aoussred (©A. Ibhi); **b.** Al Mahbes (©Woreczko Jan &Wadi); **c.** Izarzar (©Woreczko Jan &Wadi); **d.** Kheneg Ljouâd (©Stelvino).

Several more fireballs from which meteorites may have been deposited, some of which are associated with sonic phenomena, have been recorded in Morocco, but no known material that can be linked to these events has been recovered. As meteors observations are not precise, the research is not always successful, and researchers can spend weeks and months in the field without recovering the related meteorites because a nighttime fireball, after all, is visible over a much larger geographical region than that in which the meteorite physically falls.

Statistical Study of Observed Meteorite Falls in Morocco

Although statistical methods are regularly used by planetary scientists to approximate the true flux of meteorites on Earth, their distribution does not always reflects accurately specific falls. This is because some meteorite types are easier to find than others, or are degraded by weathering more quickly than others, or, especially iron meteorites, may have been collected by people in the past as they were recognized as being useful to make objects, thereby removing them from the scientific record, or, finally, valuable rare meteorite types become known to science quickly, while those of low value may never be described.

The 21 well-documented observed meteorite falls recorded in Morocco (Table **1**), consist of 3 types of samples, 17 ordinary chondrites (4 LL6, 1 EH7, 1 EH5, 1 H3-5, 1 H3-6, 4 H5, 1 CH, 2 L6, 2 L4), one carbonaceous chondrite, one eucrite unbrecciated and one Shergottite basaltic achondrites. Table **1** shows an uneven meteorite fall distribution recorded in Morocco from 1932 to 2008 (in 76 years, only 8 falls were recorded). However, between 2011 and 2020 (9 years), 12 falls were recorded, which represents a significant evolution of recorded falls. This result may be ascribed to the facts that: (i) in recent times, many hunters became specialized in collecting meteorites with an awesome skill for identifying them on the field, which, combined with scientific work, has allowed a significant increase in the number of officials falls recorded in Morocco (Ouknine, 2020); and (ii) the establishment of the Petrology, Mineralogy and Meteorites Laboratory at the Ibn Zohr University, Agadir, whose team of field researchers, headed by one of the Authors, has been directly involved in the recovery of many meteorites (Ibhi, 2012a, 2013c, 2014b, Ibhi *et al*. 2017).

The geographic distribution of Moroccan falls Fig. (**11**) shows that on a surface area of 712.5 km^2, the meteorite fall recorded rate for the 88 years from 1932 to 2020 is modest, with an average of four recoveries per 10 years, or 0.4 falls per year per 2.11 km^2. The causes for the low number of annual recorded meteorite falls are numerous and intricate. The recording of fireball observations throughout the world and a number of the potential causes of poor meteorite falls recorded rate were discussed by Rasmussen (1990). A condensed list of potential causes includes wars, the uneven distribution of people and the lack of meteorite popularization. These problems lead to the necessity of creating an academic community dedicated to the study of meteorite falls.

MOROCCAN METEORITE FINDS

General Data and Statistics

More than 50% of Morocco comprises deserts and semi-arid lands that feature favorable conditions for the prolonged preservation of meteorites (Nachit *et al.*, 2013). The official number of meteorite finds in Morocco exceeds 1740, which is a gross underestimate, because many other samples have not been submitted to the Nomenclature Committee of the Meteoritical Society because of their minor commercial value. Nearly all meteorites found in deserts fell long before humans actively sought them. The distribution of meteorite recoveries is generally concentrated in the east desert of Morocco (next to the border with Algeria), whereas few meteorites have been found around centers of population and in areas of intense agricultural activity. The majority of Moroccan meteorites in the country are held in small private collections. Meteorite finds in Morocco are divided into two groups: (i) meteorites of which the exact date and name of the locality of recovery are known and approved by the Meteorite Nomenclature Committee of the Meteoritical Society; and (ii) meteorites of which the place and date of recovery are not known exactly, which are classified with the general acronym of North West Africa (NWA) also used for not precisely allocated meteorites from Algeria, Libya, Mali, Niger, Nigeria and Mauritania (Ouknine *et al.*, 2015).

The oldest meteorite recovered in Morocco is the Douar Mghila stony meteorite that is described above, whereas the earliest well-documented recovery of large masses of iron meteorites weighing 79.9 kg occurred in 1937 near the Mrirt village (25-km north-east of Khenifra, Morocco) (Hey 1967). Another large meteorite recovery is the 100-kg mass of the Bou Azarif meteorite (a chondrite, H5) found in 2010 on the Zagora plain in east Morocco (Meteoritical Bulletin, N°100, MAPS 46). To date, the largest quantity of a single meteorite find in Morocco is that of the Agoudal iron meteorite, which amounts to several tons distributed in a few g to several hundred kg found in 2012 in the Imilchil region in Moroccan High Atlas (Ibhi and Nachit 2013, Ibhi *et al.* 2013b, 2013c, 2013d, Nachit *et al.* 2013).

Since early 2000, the recovery of meteorites in Morocco has raised markedly, largely due to discoveries in the Western Moroccan Sahara and in Eastern Morocco (Tata region), which opened many new lines of research. For climatic and geological reasons, these regions are one of the most prolific desert areas of the world for meteorite recovery (Ibhi, 2014b). The large number of meteorites found in the Western Moroccan Sahara Region does not mean that more meteorites have fallen there than anywhere else in Morocco, but it just reflects a unique physiographic environment and a sustained research effort to recover meteorites in the region. In particular, this region features a monotonic geological structure that includes rock units dating back to the Archean era, *i.e.*, more than two billion years old, and covers an area of 266,000 km². The arid to the semi-arid climate that has persisted for more than tens of thousands of years, combined with a lack of vegetation and pale rocks, has made the Western Moroccan Sahara an ideal place for the prolonged preservation and easy recognition of meteorites. Meteorites from this region now account for more than 75% of all finds in Morocco, and some places have been attributed the name of DCA, *i.e.*, places of accumulation of meteorites, by the Meteoritical Society. For example, meteorites found and classified in the area of Al Haggounia, Western Moroccan Sahara, were successively named according to the chronology of their discovery, *i.e.*, Al Haggounia 001, 002, 003, *etc.* Surprisingly, few documented discoveries exist from central and north Morocco (Ibhi, 2014b).

The 1747 distinct and authenticated meteorites with a total mass of 6175 kg currently described as originating from Morocco comprise 1674 stones (442 achondrites and 1232 chondrites), 37 irons, 24 stony-irons and 12 ungrouped meteorites. Unfortunately, most of this Moroccan material now resides in collections outside of Morocco. Table **2** shows the number of different types of meteorite finds approved and classified by the Meteorite Nomenclature Committee of the Meteoritical Society.

The total numbers and masses of known Moroccan meteorite finds recorded during the period 1937 to 2020 are shown in Figs. (**16** and **17**). After a period of stagnation from 1937 to 1997, an oscillatory growing trend in the number of recoveries occurred from 1998 to 2020, with an average of 75 new finds per year. In particular, a large increase of recoveries (131 and 149) (especially stony meteorites) was recorded respectively in 2003 and 2012, whereas the highest meteorite masses recovered in Morocco (3078 kg) occurred in 2006 (Fig. **17**). The majority of these meteorites were found in the Tata, Zogora, Errachidia, Es smara and Erfoud regions in eastern and south Morocco. Statistic data of meteorite finds in Morocco are presented in Fig. (**18**).

Table 2. Data of Moroccan meteorite finds (based on all known meteorites).

Category	Class	Sub-Classe	Type	SubType	Localized	NWA	Total
Meteorites	Undifferentiated meteorites	Chondrites	Carbonaceous		11	159	170
			Ordinary	L	29	402	431
				LL	12	192	204
				H	41	321	362
			Ungrouped		1	6	7
			Rumuruti		2	31	33
			Enstatite		2	23	25
	Differentiated meteorites	Achondrites	Ungrouped		0	5	5
			Primitive		5	35	40
			Enstatite		0	1	1
			Aubrite		1	6	7
			Angrite		0	3	3
			HED		19	214	233
			Ureilite		2	55	57
			Lunar		20	51	71
			Martian		1	36	37
		Stony-Iron Meteorites	Pallasite		2	3	5
			Mesosiderite		2	17	19
		Iron Meteorites	Iron		6	31	37
Total					156	1590	1747

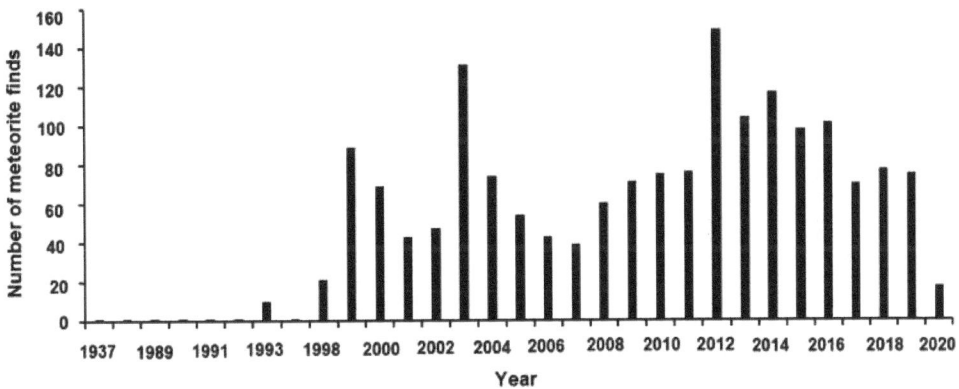

Fig. (16). Number of meteorites finds in Morocco during the period 1937-2020.

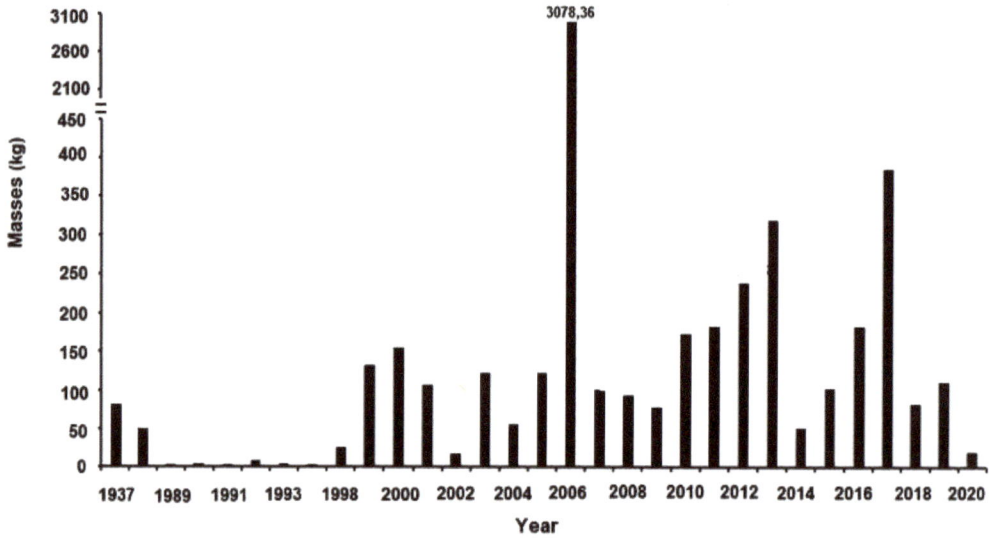

Fig. (17). Meteorites masses recovered in Morocco during the period 1937-2020.

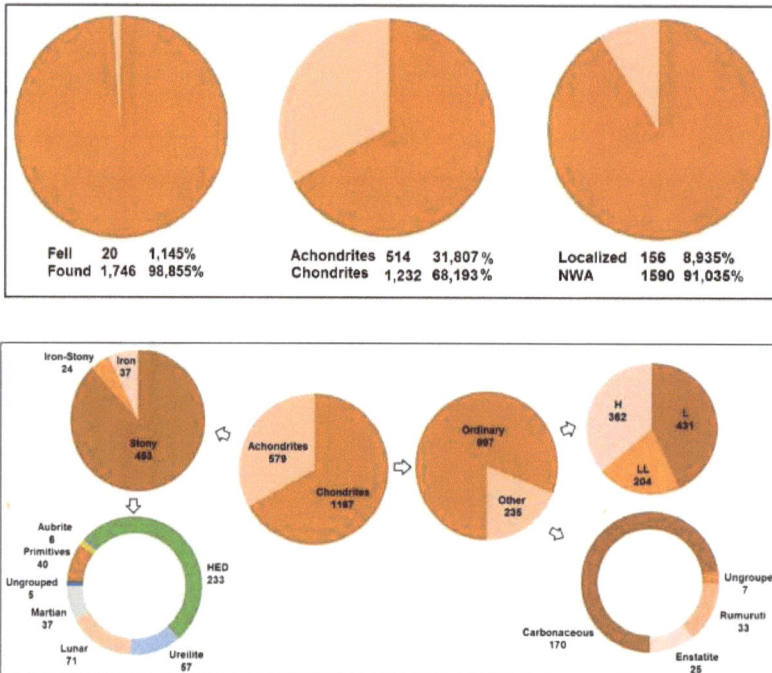

Fig. (18). Statistics of Moroccan meteorite finds.

Rare Meteorite Finds in Morocco

A few rare meteorites have been recovered in Morocco, and research on these has played an important role in extending the understanding of the early Solar System (Ibhi 2014b, 2016 and Ibhi *et al.*, 2021). One of the most remarkable of these is NWA 7325 (nicknamed "Green beauty" due to its green fusion crust), which was found in April 2012 in Bir Abbass, Boujdour (sold in a market place in Erfoud, Morocco) (Fig. **19a**). The meteorite is composed of 35 fragments with a total weight of approximately 345 g. This meteorite is the first one believed to be originated from Mercury. The oxygen isotope composition falls within the broad compositional range of ureilites and acapulcoïtes and winonaites (Irving *et al.*, 2013). The bulk major and trace element composition of the cutting dust of this meteorite measured by XRF and ICP-MS, the high Mg and low FeO contents, and the bulk Al/Si and Mg/Si ratios of this meteorite (Irving *et al.*, 2013), also in comparison with data from the Messenger mission (Weider *et al.*, 2012), suggested that it could be a plutonic rock excavated from Mercury, the age of which is estimated to be 4.56-billion-years old (Amelin *et al.*, 2013). However, a later study, based on the insufficient similarities found between the FTIR spectra of the meteorite to astronomical results of the planet surface, questioned its Mercurian origin (Weber *et al.*, 2016).

Fig. (19). Rare meteorites recovered from Morocco: **(a)** NWA 7325 "Green beauty"(© Stefan Ralew); **(b)** NWA 7034 "Black beauty" (© NASA).

Another meteorite, known as NWA 7034, and nicknamed "Black beauty", is the second oldest of the 110 meteorites retrieved on Earth and considered as originating from Mars. The meteorite was found in 2011 in Sbayta, Ghredad Sabti region, Moroccan Western Sahara, and was purchased by a meteorite dealer who sold it to a collector in USA (Ruzicka, 2015). This meteorite and its paired NWA 7533 and NWA 7475 meteorites are the first recognized Martian polymict breccia samples (Fig. **19b**). Although this meteorite originates from Mars, it does not fit into any of the meteorite categories, *i.e.*, shergottite, nakhlite, chassignite and ALH 84001, and forms a new Martian meteorite group named "Martian basaltic breccia". In particular, NWA 7034 is a volcanic breccia featuring a porphyritic appearance and consists of plagioclase and pyroxene phenocrysts up to 5 mm in diameter set in a fine-grained groundmass. Accessory minerals include chlorapatite, chromite, goethite, ilmenite, magnetite, maghemite, alkali feldspar and pyrite. Chemical analysis revealed that NWA 7034 has the highest water content ever measured in a Martian meteorite (Agee *et al.*, 2013). A team of Japanese researchers concluded that water on Mars originated about 4.4 Gyears ago (Rickman, 2019), although initial dating studies reported ages of about 2.1 Gyears. These results generated high interest in the Mars science community as this is an age not common for Martian meteorites, so subsequent studies by a variety of techniques have estimated ages of 4.48 Gyears (Bellucci *et al.*, 2015).

Other Relevant Meteorite Finds in Morocco

The meteorite NWA 12606 was found in 2018 about 30 km south-south-east of Midelt, Draa Taflalet region, Morocco, at approximate coordinates of 32°39′08.7″ N, 4°41′42.6″ W, had a total mass of 197 g and was recently classified as a main group ureilite (Gattacceca *et al.* 2020). Initial petrography for classification noted an approximate modal abundance of 90% olivine, 5% pyroxene and 5% graphite, as well as undulose extinction of olivine (Meteoritical Bulletin Data base 2020). Two thin polished sections of about 1.5 cm^2 were prepared, one to analyze petrographic and mineralogical features of silicate and carbon phases by optical microscopy and the other, not carbon coated, for SEM. The analysis of the meteorite by XRD and scanning electron microscopy–energy dispersive spectroscopy (SEM–EDS) has shown that it mainly consisted of olivine, minor pyroxene and carbon phases, possibly including carbon in the form of diamonds (Ibhi *et al.*, 2021). Carbon was usually present between mm-sized olivine and pyroxene crystals, sometimes even inside these minerals (Fig. **20**). Another study suggests that NWA 12606 has experienced intense shock metamorphism, and graphite assumed as the original carbon material might be partly transformed into diamond by a shock impact caused by meteorite collision in space (Ibhi *et al.*, 2021). Further research on this topic and analysis by other analytical techniques, such as micro-Raman (Barbaro *et al.*, 2021), is necessary to investigate this meteorite more extensively and determine the possible presence of micro diamonds and their nucleation and growth histories.

Fig. (20). Optical image of NWA 12606 ureilite (left) and backscattered electron (BSE) image (right) showing the presence of coarse-grained olivine (Ol) and pigeonite (Pig) with interstitial carbonaceous materials (Mtr), metal, dark pockets and veins of graphite-rich assemblages (Gr) and melt vein (Mv).

Two meteorites, Toufassour and Agoudal-Imilchil, one stony-iron and one iron, are associated with meteorite impact craters. The existence of a small, isolated crater of 3-m diameter and 1.25 m-depth of meteoritic origin in the Toufassour region (Tata, Morocco) was brought to our attention late in 2006 by a nomad, who found many mesosiderite fragments in the vicinity. The crater is located at 29°39.135'N, 07°44.958'W, at an altitude of 478 m, in an arid alluvial fan, approximately 15 km from the foot of the Bani Mountains. While impact breccias were not found at Toufassour, meteoritic material was found on the south and northeast outer part of the pit rim. The presence of the mesosiderite blocs and debris throughout the pit confirmed the hypothesis that this relatively small pit was caused by a penetrative, rather than the explosive, the impact of a meteoritic body of moderate mass and high density, moving at a relatively low velocity (Ibhi & Darbali, 2011, 2012).

The NWA 1827, 1879, 1882, 1912 and 1951 mesosiderite meteorite fragments ranging in size from a few g to more than 5 kg, were recovered in the Toufassour region in 2003. After the easy recovery of about 300 kg from the surface of a 50-m by 100-m area along the southwest to the northeast axis (N30) (Ibhi *et al.*, 2008), the Toufassour area has been the focus of extensive meteorite collecting efforts that continue even today. Various studies by the LPMM Laboratory of the University Ibn Zohr (Morocco) and the Magie Laboratory of the University of Paris VI (France) have confirmed the meteoric origin of Toufassour recoveries. The fragments feature the absence of fusion crust and a slightly glossy surface. Metal nodules protrude here and there, and larger silicate inclusions (Fig. **21A**) are visible as greenish spots (Ibhi *et al.*, 2008). The modal composition is relatively homogeneous, and the modal contents of orthopyroxenes are higher than those of plagioclases, thus this mesosiderite is of type B. The silicate matrix is characterized by significant degrees of metamorphism (intergranular, type 4).

The fragments are mainly composed of metal (free or nodule) and silicates with Fe-Ni metals forming irregular particles of kamacite with round or lamellar taenite, inclusions of schreibersite and rare troilite, and silicate minerals consisting of plagioclase with poikilitic texture, pyroxenes and olivine. Clinopyroxene crystals are smaller than those of orthopyroxene and olivine (Fig. **21B**). In particular, the composition (% vol) of silicates is: pyroxene (51.6%), plagioclase (33.6%), silica (5.89%), Ca-phosphate (0.48%), chromite and rare ilmenite (0.36%), iron hydroxide (1.76%), troilite (0.27%) (Ouknine *et al.*, 2019; Ouknine, 2020). An impact crater of the size of the Toufassour depression is expected to belong to a swarm, and, therefore, should be surrounded by a cluster of pits that would result from the fragmentation of brittle cosmic objects during their flight through the atmosphere. Within the region, plans are underway to conduct additional magnetic and geological studies of the pits and their vicinities.

Fig. (21). Toufassour mesosiderite. MEB (Ol: olivine, Opx: orthopyroxene, Pl: plagioclase, Met: metal, Chr: chromite, Ms: maskelynite and Slf: sulfides) (Ouknine, 2020).

The first fragment of the Agoudal-Imilchil meteorite was collected by a nomad in 2007. In 2012, many pieces of this meteorite arrived at the Laboratory of Ibn Zohr University for analysis. Immediately a systematic search for meteorites was started in collaboration with professional prospectors and merchants to define the direction of the strewnfield. The members of the laboratory collected many fragments of the meteorite (Fig. **22**) at various GPS coordinates: Tasrafet (32°10'56.4"N, 05°52'59.5"W), Akdim (32°04'00.5"N; 05°28'57.3"W), Bouzmou (32°05'35.9"N, 05°32'12.0"W), Tighdouine (32°13'04.2"N, 05°26'42.3"W) and Agoudal, (31°59'08.0"N, 05°31'00.6"W). The fragment was found nearby Tasraft, a small village situated at about 30 km west of Agoudal, and all other specimens collected in the region had the same mineralogical characteristics of an IIAB iron meteorite (Ibhi *et al.*, 2013b; Ibhi *et al.*, 2013c). The bulk chemical analysis of the samples (Ruzicka *et al.*, 2015) revealed an average composition of

6%wt Ni, 4 mg/g Co, 60 µg/g Ga and < 0.05 µg/g Ir. SEM analysis revealed a coarse pattern of irregular interlocking kamacite grains with some sub-boundaries. The samples do not show Widmanstätten figures and schreibersites, and (Fe, Ni)$_3$P, was present abundantly as mm-sized skeletal crystals at the centers of kamacite crystals. Schreibersite textures are well known to be the result of high-speed collisions occurring in the asteroid belt (Britvin *et al.*, 2002). Over the course of six years, samples weighing from 100 g to more than 196 kg were collected, and more than a ton of specimens were gathered, although other reports claim that a much higher amount (up to 2 tons) has been found. Due to the continuing intense exploration, specimens are still found. The mapping of the locations where the fragments of the meteorite were found suggests that the fireball exploded, releasing thousands of fragments that were scattered on a strewnfield about 38-km long in the north-south direction (Nachit *et al.*, 2013).

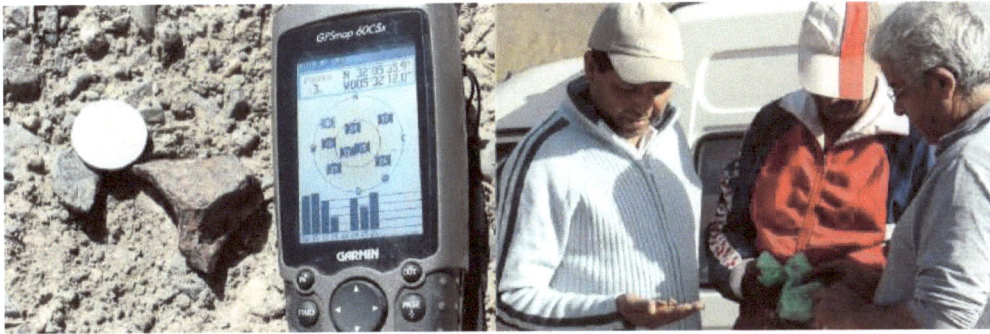

Fig. (22). Meteorite fragments of Agoudal (Imilchil region) (© A. Ibhi).

The Agoudal iron meteorite field is the only verified impact site in Morocco where the remains of a number of impact structures have been found. The impact crater is located about 20 km south of Imilchil (31°59'12.7''N, 5°30'57.3''W), Central High Atlas, and was recognized by Sadilenko *et al.* (2013) thanks to the presence of shatter cones and breccias in ravines that cut the sedimentary formations of this place. The mission we conducted on the terrain showed that the location of the crater corresponds to the flank of a syncline with a dip of 25° to the north. Furthermore, the whole north-western part of the synclinal flank, where the crater is located, is washed out by a big valley of 300-m width. Very beautiful shatter cones (Fig. **23**) have been found in samples of Jurassic limestone wrapped in quaternary deposits, predominantly at the flanks of the dry rivers that cut the syncline. Resistivity data were collected in the central uplift of the impact structure by using a six-current electrode dipole-dipole configuration (Tong *et al.*, 2010). The profile showed a large conductive anomaly with low resistivity values centered in the profile (250 Ωxm) with thicknesses up to 70 m in the center of the investigated area. This anomaly is lenticular in shape, 300-m long, and characterized by sudden variations of resistivity between 140 to 1000 Ωxm. The profile also shows that the fall of the meteorite probably occurred in the NE part of the crater, where the highly fractured rocks have the lowest resistivity values (100 Ωxm). The tomographic profiles have shown the presence of an impact crater, 300-m in diameter (Fig. **24**), almost completely covered by a block of limestone rocks, 220-m long and 40-m deep (Ibhi & Nachit, 2013; Ibhi *et al.*, 2013d). Furthermore, the obliquity of the impact probably produced a crater that was initially rimless. The northern part of the synclinal flank, where the crater is located, is washed out by the Tamegzarout valley, which makes it difficult to determine the exact form of the crater that should, theoretically, be slightly stretched in the NE-SW direction. The supposed original form is now roughly bounded by rivers that are actually faults propagating to the east, west and south. Finally, the freshness of the meteorites found to show that the age of the crater is relatively recent, possibly the quaternary age.

Fig. (23). The Agoudal impact crater (left) and shatter cones in the breccias (right). (© A. Ibhi).

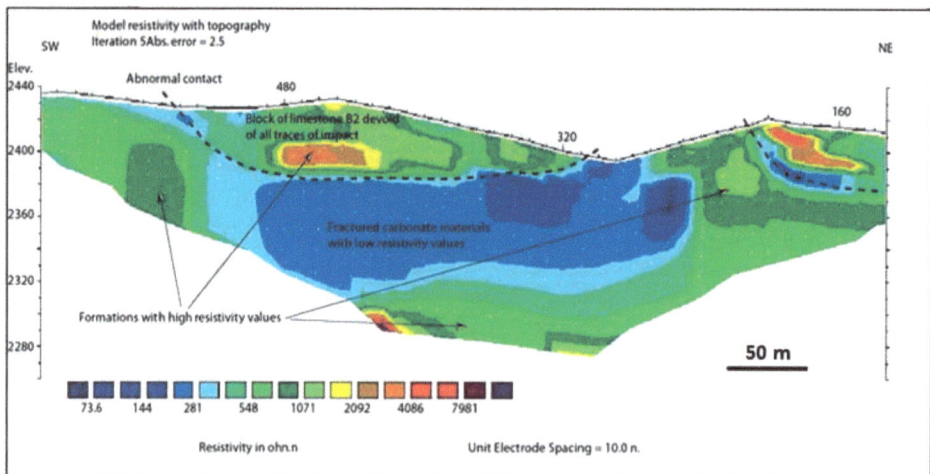

Fig. (24). 2D-electrical resistivity tomography model of the impact structure.

Fragments of a new iron meteorite have been found and collected at Oglat Sidi Ali (Eastern Highlands of Morocco), about 20 km north of the Maatarka region (coordinates 33°31'45" N 02°37'01" W). Since 1998 several thousands of small (< 10 kg), medium (10 to 50 kg) and large-sized (> 50 kg) meteorite fragments were recovered in this area with a total estimated mass of more than 800 kg spread across the NE–SW oriented, 20-km long and 5-km wide strewnfield (Nachit *et al.*, 2022). Analytical data performed on an etched and polished meteorite section suggested a classification as an ungrouped iron meteorite. Part of these data was used to classify the meteorite fragments upon their submission to the Nomenclature Committee of the Meteoritical Society for official approval (Bouvier *et al.*, 2017).

Successively, in the years 2013, 2014, 2015 and 2017, four fieldwork expeditions were conducted at Oglat Sidi Ali by some of the authors with the aid of a metal detector and the collaboration of dozens of nomads. A total of 72 iron meteorite fragments were recovered for a total weight of approximately 600 g. The type specimens, consisting of 3 small fragments totally weighing 30 g, are deposited at the Museo di Storia Naturale, Università di Firenze, Italy; two pieces totally weighing 14.5 g and including an etched and polished mount are deposited at the Museo di Scienze Planetarie, Prato, Italy; 42 pieces totally weighing 195 g are conserved at the Ibn Zohr University, Morocco; and one sample was submitted to the Meteoritical Society and approved as a new meteorite in 2015 with the name Oglat Sidi Ali (Moggi Cecchi *et al.*, 2015; Bouvier *et al.*, 2017). Most recovered specimens exhibit various shapes and feature smooth surfaces covered by cavities (regmaglypts) that were formed by ablation due to the frictional heating that occurred during the atmospheric passage of the meteorite (Fig. **25**). Low-magnification optical microscope analyses on a cut and etched surface allowed the examination of the internal texture, that displays a "plessitic octahedrite" texture consisting of elongated spindles of kamacite set in a groundmass of a fine-grained plessitic intergrowth of kamacite and taenite.

Fig. (25). Physical characteristics of individual samples of the Oglat Sidi Ali iron meteorite.

More detailed information was obtained by means of electron backscattered diffraction (EBSD) that allowed us to distinguish the body-centered cubic (bcc) alpha phase (kamacite) from the face-centered cubic (fcc) gamma phase (taenite) of Fe-Ni alloys (Nachit *et al.*, 2022). Taenite featured a mean Ni content of 28.8 wt %, with a maximum of 42.6 wt % for tetrataenite grains; kamacite showed a mean Ni content of 6.3 wt %; and the plessitic area displayed a mean Ni content of 15.9 wt %. The bulk analyses of major, minor and traces elements, which are requested for classifying iron meteorites (Grady *et al.*, 2014), resulted in high Ni (140.7) and Co (11.7) contents (both of the order of mg/g), as well as high Cu (249 µg/g), Ga (79.1 µg/g), Ru (53.22 µg/g), Pd (8.79 µg/g) and Pt (43.9 µg/g) contents and relatively low W (5.79 µg/g), Ir (2.5 µg/g), As (9.99 µg/g) and Re (0.28 µg/g) contents. Ge and Au concentrations could not be determined by ICP-MS with a high degree of certainty, however, an estimate of the overall Ge content (2256 µg/g) was obtained by means of PIXE analysis.

Twelve groups of iron Oglat Sidi Ali meteorites are currently recognized and designated by Roman numerals (I, II, III, IV) and letters A through F according to the concentrations of selected siderophile trace elements (such as Ga and Ir) plotted against the overall Ni content on a logarithmic plot (D'Orazio, 2020). Although the meteorites texture shows characteristics similar to those of other plessitic octahedrites, the high Ni, Ga, Ru, Pd and Pt contents, as well as the relatively low Ir contents, are outside the limits for the IIC or IIF groups (Fig. **26**), thus the classification as an ungrouped plessitic octahedrite iron meteorite was confirmed (Moggi Cecchi *et al.* 2015).

A comparison of this meteorite with other ungrouped iron meteorites purchased between 2001 and 2016 in various cities of North-East Morocco showed apparently similar mineralogy, geochemistry and textural features, which suggested a common origin from a single extraterrestrial body. Plotting the Ga and Ir contents of the Oglat Sidi Ali meteorite and those of other ungrouped iron meteorites, namely NWA 859 (also known as Taza), NWA 11010 (Bouvier *et al.*, 2017b) and NWA 7335, resulted in strong geochemical affinities, although differences are evidenced for NWA 7335 (Fig. **26**). Furthermore, the comparison of literature data for a set of siderophile minor and trace elements in these meteorites showed that, with the exception of the lower As the content of the Oglat Sidi Ali meteorite, the contents of the other elements are similar, which confirmed a possible common genetic origin for all these ungrouped plessitic iron meteorites.

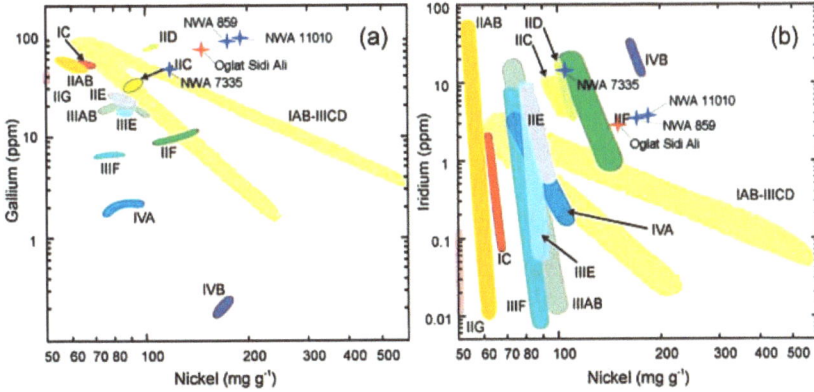

Fig. (26). Logarithmic plot of the bulk amounts of Ga and Ir (µg/g) versus Ni (mg/g) of the Oglat Sidi Ali meteorite and some NWA ungrouped irons and literature data for various iron meteorite groups (modified after Grady *et al.*, 2014).

In order to test a possible pairing among these meteorites, their compositional data for eight main siderophile elements (Ni, Co, Ge, Cu, Ga, W, Ir and Pt) were compared with those of the ungrouped iron Butler meteorite (Fig. **27**). The NWA 859 and NWA 11010 have compositions quite similar to that of Oglat Sidi Ali, while NWA 7335 has marked differences in the Ni, Co, Ga, W and Pt contents. As it concerns the textural features, the Oglat Sidi Ali meteorite shows a plessitic structure similar to that of NWA 859 and NWA 11010, at both the hand specimen scale and at the microscopic scale, with spindles of kamacite separated by plessite fields, similar to those already described by Wasson (2011) for the Butler ungrouped iron meteorite. Thus, based on presently available chemical and textural evidences it seems probable that NWA 859 and NWA 11010 are paired with Oglat Sidi Ali, while NWA 7335 has to be considered a separate find.

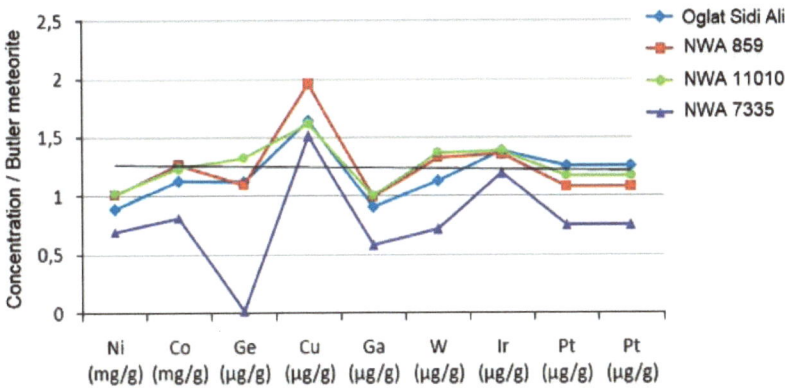

Fig. (27). Siderophile element abundances of selected NWA ungrouped iron meteorites normalized to those of the Butler meteorite (Goldstein, 1966).

On February 11, 2023, the Nomenclature Committee (NomCom) of the Meteoritical Society officially published the declaration of an interesting meteorite collected in Morocco. This is "Smara 002" (which will be published in Meteoritical Bulletin, no. 111, in preparation for 2023), whose classification and scientific work were carried out by the members of the laboratory of the Dipartimento di Scienze della Terra, Università di Firenze, Italy and at the Scientific Research center of the Ibn Zohr University, Agadir Morocco and at the Centro di Servizi di Microscopia Elettronica e Microanalisi (MEMA) of the University of Firenze, Italy (Fig. **28**).

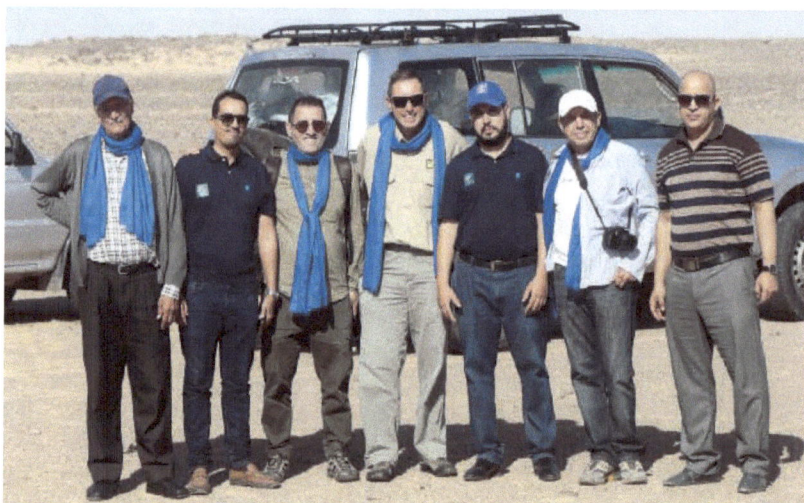

Fig. (28). The "Smara 002" discovery fieldwork team.

A single piece of about 155 grams (Fig. **29a**) and several small fragments, a total weighing 805 grams, were found on 15 March 2021 by Mouloud Bachikh and a group of friends during a meteorite search in the El-Aasli area near Jabal Kaur Al-Bared, west of Es Smara, Morocco (GPS coordinates 26°40.35' N, 11°46.16' W). The meteorite is partially covered with fresh fusion crust. This meteorite shows a chondritic texture with relatively large, well-formed porphyritic olivine-pyroxene (POP) and porphyritic olivine (PO) chondrules, ranging in size from 400 to 1500 µm and set in a fine-grained, dominant (75 vol.%) silicate matrix (Fig. **29b**). Large CAIs ranging in size from 400 to 800 µm and consisting of melilite and spinel are present. The matrix is mainly consisting of Fa-rich olivine with scattered iron oxides. The meteorite has moderate weathering and a low-impact state. The magnetic susceptibility measurement performed on two fragments of the meteorite showed that Log X (10-9 m3/kg) is about 4.62. This value corresponds well to the confidence interval of the CV meteorites in the alignment chart given by Folco *et al.* (2006).

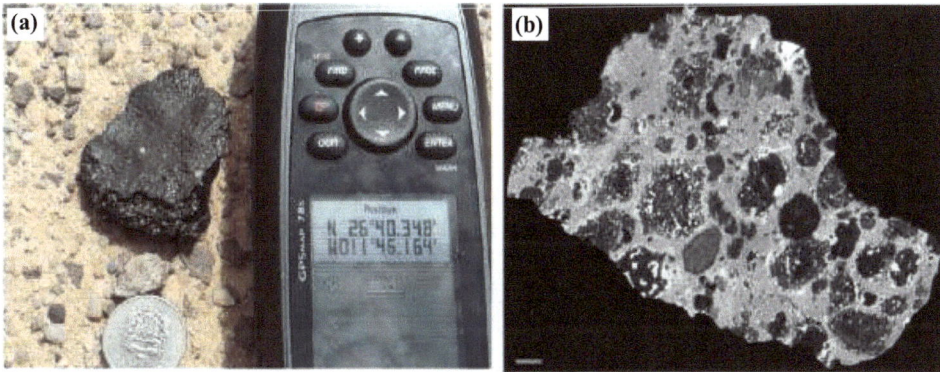

Fig. (29). Image of one of a fragment of the meteorite Smara 002 on the site of recovery **(a)**. SEM-BSE photomosaic image of the meteorite displaying the general texture; the scale bar is 500 μm **(b)**.

According to textural and compositional data, as well as to the measured magnetic susceptibility, the meteorite was classified as an unequilibrated, Carbonaceous chondrite belonging to the CV group (Moggi-Cecchi *et al.*, 2023). The marked compositional inhomogeneity and the well-formed chondrules suggest a petrologic type 3. The presence of pentlandite and magnetite suggests that the meteorite belongs to the oxidized-Bali (ox-B) subgroup (Braukmüller *et al.*, 2018). The type specimens, weighing 22 g, and one thin section are on deposit at the Museo di Storia Naturale-Sistema Museale di Ateneo of the University of Firenze, Italy (Inv. # I-3710). One polished thin section is on deposit at the University Museum of Meteorites (UMM), Agadir, Morocco. Bachikh Mouloud holds the main mass.

CONCLUSION

The importance of Moroccan meteorites in basic research is becoming more and more significant in a variety of fields. New groupings of meteorites are being found, which is expanding our understanding of the early Solar System. The official number of meteorites coming from Morocco exceeds 1700, which is a gross underestimate as the find places and coordinates of a significant number of meteorites collected in Morocco are not precisely documented, and were classified generically as NWA meteorites.

The astronomical illustrations carved into rocks attest that ancient humans were interested in observing meteors and meteorites. Three rocks art in particular, which were discovered close to Ida OuKazzou (about 100 km north of Agadir), would indicate that prehistoric Moroccans saw meteorite falls.

Of the over 1700 official Moroccan meteorites, only 21 well-documented meteorite falls have been recorded by the end of August 2020, which were all observed by eyewitnesses as they moved through the atmosphere and touched the ground. Unfortunately, most of this material now resides in collections outside the country. The first one, Douar Mghila (LL6), was observed in 1932, and the last one, called Tiglit, in 2021. The meteorite fall recovery rate during the past 88 years is low on average, *i.e.*, approximately four falls recorded every 10 years.

A total of 1747 certified meteorite finds with a total mass of 6175 kg have been identified as originating from Morocco. The Es Smara, Zogora, Tata, Erfoud and Errachidia areas in eastern and southern Morocco are the localities where most of these meteorites were recovered. The Agoudal iron meteorite, which was discovered in 2012 in the Imilchil area of the Moroccan High Atlas, comprises many tons of fragments, each weighing from a few grams to several hundred kilograms, and remains the largest meteorite fall in Morocco.

A number of rare and unknown meteorites have been collected in Morocco, of which the most significant is the NWA 7325 meteorite nicknamed "Green beauty," which is the first meteorite believed to be originated from Mercury. Another meteorite, known as NWA 7034 and nicknamed "Black beauty", is the second oldest of 110 named meteorites originating from Mars and retrieved on Earth. Two meteorites have been found in Morocco that show impact craters, *i.e.*, the mesosiderite found in 2003 in the Toufassour region and the Agoudal meteorite, of which the first fragment was found by a nomad in 2007.

REFERENCES

Agee, C.B., Wilson, N.V., McCubbin, F.M., Ziegler, K., Polyak, V.J., Sharp, Z.D., Asmerom, Y., Nunn, M.H., Shaheen, R., Thiemens, M.H., Steele, A., Fogel, M.L., Bowden, R., Glamoclija, M., Zhang, Z., Elardo, S.M. (2013). Unique meteorite from early Amazonian Mars: water-rich basaltic breccia Northwest Africa 7034. *Science, 339*(6121), 780-785.
[http://dx.doi.org/10.1126/science.1228858] [PMID: 23287721]

Ait Kadi, M. Bulletin d'information de l'académie hassan ii des sciences et techniques. (2012). *12*, 106-108.

Amelin, Y., Koefoed, P., Iizuka, T., Irving, A.J. (2013). U-Pb age of ungrouped achondrite NWA 7325. *In 48th Annual Meteoritical Society Meeting, #5165 (abstr.).*

Bailey, M.E. (1995). Recent results in cometary astronomy: Implications for the ancient sky. *Vistas Astron., 39*(4), 647-671.
[http://dx.doi.org/10.1016/0083-6656(95)00013-5]

Barbaro, A., Nestola, F., Pittarello, L., Ferrière, L., Murri, M., Litasov, K.D., Christ, O., Alvaro, M., Domeneghetti, M.C. (2022). Characterization of carbon phases in Yamato 74123 ureilite to constrain the meteorite shock history. *Am. Mineral., 107*(3), 377-384.
[http://dx.doi.org/10.2138/am-2021-7856]

Bednarik, R.G. (2008). Cupules. *Rock Art Res., 2008*(25), 61-100.

Beech, M. (2003). *J. R. Astron. Soc. Can., 97*, 71-77.

Bellucci, J.J., Nemchin, A.A., Whitehouse, M.J., Humayun, M., Hewins, R., Zanda, B. (2015). Pb-isotopic

evidence for an early, enriched crust on Mars. *Earth Planet. Sci. Lett., 410*, 34-41.
[http://dx.doi.org/10.1016/j.epsl.2014.11.018]

Bouvier, A., Gattacceca, J., Agee, C., Grossman, J., Metzler, K. (2017). The Meteoritical Bulletin, No. 104. *Meteorit. Planet. Sci., 52*(10), 2284.
[http://dx.doi.org/10.1111/maps.12930]

Braukmüller, N., Wombacher, F., Hezel, D.C., Escoube, R., Münker, C. (2018). The chemical composition of carbonaceous chondrites: Implications for volatile element depletion, complementarity and alteration. *Geochim. Cosmochim. Acta, 239*, 17-48.
[http://dx.doi.org/10.1016/j.gca.2018.07.023]

Britvin, S.N., Rudashevsky, N.S., Krivovichev, S.V., Burns, P.C., Polekhovsky, Y.S. (2002). Allabogdanite, (Fe,Ni) $_2$ P, a new mineral from the Onello meteorite: The occurrence and crystal structure. *Am. Mineral., 87*(8-9), 1245-1249.
[http://dx.doi.org/10.2138/am-2002-8-924]

Buchner, E., Schmieder, M., Kurat, G., Brandstätter, F., Kramar, U., Ntaflos, T., Kröchert, J. (2012). Buddha from space-An ancient object of art made of a Chinga iron meteorite fragment. *Meteorit. Planet. Sci., 47*(9), 1491-1501.
[http://dx.doi.org/10.1111/j.1945-5100.2012.01409.x]

Chen, K., Wang, Y., Liu, Y., Mei, J., Jiang, T. (2018). Meteoritic origin and manufacturing process of iron blades in two Bronze Age bimetallic objects from China. *J. Cult. Herit., 30*, 45-50.
[http://dx.doi.org/10.1016/j.culher.2017.10.004]

Coimbra, F.A. (2007). Comets and meteors in rock art: evidences and possibilities. *13th SEAC Conference Proceedings, Isili.,* 250-256.

Coimbra, F.A. (2009). When open air carved rocks become sanctuaries: methodological criteria for a classification. In: Djindjian, F., Oosterbeek, L., (Eds.), *Proceedings of XV IUPPS Congress, Archaeopress,* Oxford 200999-104.

Coimbra, F.A. (2010). The sky on the rocks: cometary images in rock art. Actas do Congresso Internacional de Arte Rupestre "Global Art 2009", Brasil: Serra da Capivara, 635-646.

Comelli, D., D'orazio, M., Folco, L., El-Halwagy, M., Frizzi, T., Alberti, R., Capogrosso, V., Elnaggar, A., Hassan, H., Nevin, A., Porcelli, F., Rashed, M.G., Valentini, G. (2016). The meteoritic origin of Tutankhamun's iron dagger blade. *Meteorit. Planet. Sci., 51*(7), 1301-1309.
[http://dx.doi.org/10.1111/maps.12664]

D'Orazio, M. (2020). Petrography, geochemistry and classification of ten new iron meteorites from Northwest Africa and Chile. *Period. Mineral., 89*, 3-18.
[http://dx.doi.org/10.2451/2020PM855]

Figueiredo, D.A., Coimbra, F.A., Monteiro, D.C., Ribeiro, D.N. (2017). Preliminary analysis of the rock art from Buracas da Serra, Alvaiázere (Portugal). Estudiopreliminardelarte rupestre de la Sierra de Buracas, Alvaiázere (Portugal). *Revista Cuadernos De Arte Prehistorico, 4*, 127-140.

Folco, L., Rochette, P., Gattacceca, J., Perchiazzi, N. *In situ* identification, pairing, and classification of meteorites from Antarctica through magnetic susceptibility measurements. (2006). *Meteorit. Planet. Sci., 41*(3), 343-353.
[http://dx.doi.org/10.1111/j.1945-5100.2006.tb00467.x]

Gattacceca, J., Bouvier, A., Grossman, J., Metzler, K., Uehara, M. (2019). The Meteoritical Bulletin, No. 106. *Meteorit. Planet. Sci., 54*(2), 469-471.
[http://dx.doi.org/10.1111/maps.13215]

Gattacceca, J., Mccubbin, F.M., Bouvier, A., Grossman, J.N. (2020). The Meteoritical Bulletin, no. 108. *Meteorit. Planet. Sci., 55*.

Gattacceca, J., McCubbin, F.M., Grossman, J., Bouvier, A., Bullock, E., Chennaoui Aoudjehane, H., Debaille, V., D'Orazio, M., Komatsu, M., Miao, B., Schrader, D.L., Timothy Jull, A.J. (2021). The

Meteoritical Bulletin, No. 109. *Meteorit. Planet. Sci.,* *56*(8), 1626-1630. [http://dx.doi.org/10.1111/maps.13714]

Gattacceca, J., Mccubbin, F.M., Bouvier, A., Grossman, J. (2020). The Meteoritical Bulletin, No. 107. *Meteorit. Planet. Sci.,* *55*(2), 460-462. [http://dx.doi.org/10.1111/maps.13440]

Goldstein, J.I. (1966). Butler, missouri: an unusual iron meteorite. *Science,* *153*(3739), 975-976. [http://dx.doi.org/10.1126/science.153.3739.975] [PMID: 17837245]

Grady, M., Pratesi, G., Moggi Cecchi, V. (2014). *Atlas of Meteorites..* Cambridge University Press. 384.

Grady, M.M. Catalogue of meteorites. cambridge university press: cambridge; new york. (2000). Madrid, Cape Town: Oakleigh. 689.

Grossman, J.N., Zipfel, J. (2001). Meteoritical Bulletin, No. 85. *Meteorit. Planet. Sci.,* *36*(S9), A293-A322. [http://dx.doi.org/10.1111/j.1945-5100.2001.tb01542.x]

Hassiba, R., Cieslinski, G.B., Chance, B., Abdulla Al-Naimi, F., Pilant, M., Rowe, M.W. (2012). Determining the age of Qatari Jabal Jassasiyah Petroglyphs. *QScience Connect,* *2012*(1), 2-16. [http://dx.doi.org/10.5339/connect.2012.4]

Hey, M.H. (1967). Meteoritical Bulletin 40. *Meteoritics,* *5*, 85-109.

Ibhi, A. (2012). The martian meteorite of tissint (tata, morocco) preliminary scientific study and reports of the saharan nomads. *Meteorite,* *18*(2), 9-12.

Ibhi, A. (2012). Une martienne tombée au Maroc. *Astronomie-Bulletin de la Société Astronomique de France,* *48*, 52-55.

Ibhi, A. (2012). New mars meteorite fall in morocco: collecting observations and spatial distribution in the strewnfield. *International Meteor Conference (IMC).*20-23 SeptemberLa Palma, Canary Islands, Spain

Ibhi, A. (2013). New mars meteorite fall in morocco: collecting observations and spatial distribution in the strewnfield. *Meteoroids,* *2013*, 26-30.

Ibhi, A. (2013). New mars meteorite fall in marocco: final strewn field. international letters of chemistry. *Physics and Astronomy,* *11*, 20-25.

Ibhi, A. (2013). Meteors and meteorite falls in morocco. international letters of chemistry. *Physics and Astronomy,* *17*, 28-35.

Ibhi, A. (2014). Tighert: A new eucrite meteorite fall from Morocco. *Proceedings of the IMC, Proceedings of the International Meteor Conference.*18-21 SeptemberGiron, France

Ibhi, A. (2014). Morocco meteorite falls and finds: some statistics. International Letters of Chemistry. *Physics and Astronomy,* *1*, 18-24.

Ibhi, A. (2016). *Météorites, Perles du Désert Marocain..* Edition MUM.

Ibhi, A. (2019). Large meteorite (H4-5) exploded over the region of Zag (Morocco). *Meteor News,* *4*, 1-4.

Ibhi, A. (2022). Exploding meteor over Tiglit (Morocco) with fragments of an interesting meteorite collected. *Meteor News,* *7*(2), 69-72.

Ibhi, A., Nachit, H., Jambon, A., Boudouma, O., Badia, D. (2008). Meteoritical Bulletin, N°94. *Meteorit. Planet. Sci.,* *43*, 1551-1588.

Ibhi, A., Darbali, M. (2011). The discovery of impact crater in southeast Morocco. *2nd Planetary Crater Consortium Meeting.*Sept. 14-16Flagstaff, USA

Ibhi, A., Darbali, M. Les météorites du Maroc: Les mésosidérites du Maroc et la structure d'impact associée. (2012).

Ibhi, A., Nachit, H. (2013). The enigmatic shatter cones of Agoudal (Imilchil, Morocco). *4th Planetary Crater Consortium Meeting*August 14-16US Geological Survey.Flagstaff, USA:

Ibhi, A., Nachit, H. (2013). Tissint Meteorite: New Mars Meteorite fall in Morocco. *J. Mater. Environ. Sci.,* *4*(2), 293-298. a

Ibhi, A., Nachit, H., Abia, E.H., Touchnt, A.A., Vaccaro, C. (2013). Isli and Tislit: The First Dual Impact Crater Discovered in Morocco. *International Journal of Astronomy and Astrophysics, 3*(2), 1-4. [http://dx.doi.org/10.4236/ijaa.2013.32A001]

Ibhi, A., Nachit, H., Abia, El. (2013). Discovery of the double impact crater in the Imilchil region, Morocco. *European Planetary Science Congress, 8*, 741.

Ibhi, A., Nachit, H., Abia, El. (2013). Discovery of the double impact crater in the Imilchil region, Morocco. *4th Planetary Crater Consortium Meeting. August,* US Geological Survey.Flagstaff, USA1-2. d

Ibhi, A., Khiri, F., Ouknine, L. (2017). New meteorite fall in the region of Igdi (Tata, Morocco). *Proceedings of the IMC.,* Petnica.Serbia: 1-3.

Ibhi, A., Khiri, F., Ouknine, L., Lemjidi, A., Asmahri, El. (2018). Rock Art and Archaeoastronomy in Morocco: Preliminary Observations. *Archaeoastronomy and Ancient Technologies., 6*(2), 44-54.

Ibhi, A., Khiri, F., Ouknine, L., Lemjidi, A., Asmahri, El. (2019). The discovery of mysterious petroglyphs suggests that a meteor has been observed in ancient times in Morocco. *MeteorNews, 4*(1), 1-3.

Ibhi, A., Khiri, F., Ouknine, L., Ait Touchnt, A., Capitelli, F., De Pascale, O., Senesi, S. (2021). A new carbon-rich ureilite from Morocco. Rendiconti Lincei. Scienze Fisiche e Naturali. *Scienze Fisiche e Naturali, 32*(4), 709-713.

Ibhi, A., Ouknine, L., Khiri, F., Ait Touchnt, A., Nachit, H., De Pascale, O., Senesi, G.S. (2022). The meteoritic origin of morocco iron dagger blades. *Heritage, 5*(3), 1395-1400. [http://dx.doi.org/10.3390/heritage5030072]

Iqbal, N., Vahia, M.N., Ahmad, A., Masood, T. (2010). The prehistoric meteor shower recorded on a Palaeolithic Rock. *NRIAG Journal of Astronomy and Astrophysics (Egypt), 2010*(Special Issue), 469-475.

Iqbal, N., Vahia, M.N., Masood, T., Ahmad, A. (2009). Some early astronomical sites in the Kashmir region. *Journal of Astronomical History and Heritage, 12*(1), 61-65. [http://dx.doi.org/10.3724/SP.J.1440-2807.2009.01.05]

Irving, A.J., Kuehner, S.M., Bunch, T.E., Ziegler, K., Chen, G., Herd, C.D.K., Conrey, R.M., Ralew, S. (2013). Petrology and oxygen isotopic composition of brachinite-like achondrites Northwest Africa 7399 and Northwest Africa 7605, and evidence for late-stage methane-triggered reduction. *44th Lunar and Planet. Sci. Conf.*

Jambon, A. (2017). Bronze Age iron: Meteoritic or not? A chemical strategy. *J. Archaeol. Sci., 88*, 47-53. [http://dx.doi.org/10.1016/j.jas.2017.09.008]

Khiri, F., Ibhi, A., Ouknine, L. (2015). Temporal and spatial distribution of meteorites falls in Africa. In: Rault, J.-L., Roggemans, P., (Eds.), *Proceedings of the International Meteor Conference,* Mistelbach, Austria27-30 August 2015International Meteor Organization.978-2-87355-029-5117-119.

Khiri, F., Ibhi, A., Saint-Gerant, T., Medjkane, M., Ouknine, L. (2017). Meteorite falls in Africa. *J. Afr. Earth Sci., 134*, 644-657. [http://dx.doi.org/10.1016/j.jafrearsci.2017.07.022]

Köehler, U. (2002). Meteors and comets in ancient mexico, catastrophic events and mass extinctions: impacts and beyond, christian koeberl, Kenneth G. MacLeod Geological Society of America. [http://dx.doi.org/10.1130/0-8137-2356-6.1]

Kotowiecki, A. (2004). Artifacts in Polish collections made of meteoritic iron. *Meteorit. Planet. Sci., 39*(S8) (Suppl.), A151-A156. [http://dx.doi.org/10.1111/j.1945-5100.2004.tb00351.x]

Lacroix, M.A. (1933). *C. R. Hebd. Seances Acad. Sci., 197*, 368-369.

Millot, G., Nahon, D., Paquet, H., Ruellan, A., Tardy, Y. (1977). L'épigénie calcaire des roches silicatées dans les encroûtements carbonatés en pays subaride. *Antiatlas, Maroc. Sciences Géologiques, bulletins et mémoires, 30*(3), 129-152.

Moggi Cecchi, V., Caporali, S., Pratesi, G., Nachit, H., Herd, C.D.K., Chen, G. (2015). Compositional and textural data of a new ungrouped iron meteorite from OglatSidi Ali, Morocco. *Eur. Planeta. Sci. Congr. Abst., 10*, 312-314.

Moggi-Cecchi, V., Ibhi, A., Nachit, H., Pratesi, G., Senesi, G.S. (2023). A new carbonaceous chondrite from the Es Smara region, Western Sahara: compositional and textural data. *54th Lunar and Planetary Science Conference (LPSC)*.March 13–17Houston, Texas, USA

Nachit, H., Ibhi, A., Vaccaro, C. (2013). The Imilchil meteorite strewed field and Isli-Agoudal craters. *International Letters of Chemistry, Physics and Astronomy 11.,* 65-71.

Nachit, H., Ibhi, A., En-nasiry, M., Moggi Cecchi, V., Pratesi, G., Herd, C.D.K., Senesi, G.S. (2022). Minerochemical and microtextural study of the ungrouped iron meteorite oglat sidi ali, eastern highlands, morocco, and geomorphological characterization of its strewnfield. *Minerals (Basel), 12*(11), 1470. [http://dx.doi.org/10.3390/min12111470]

Olivera, C., Silva, C.M. (2010). Moon, spring and Large Stones. In: Calado, D., Baldia, M., Boulanger, M., (Eds.), *Proceedings of XV IUPPS Congress, Archaeopress,* Oxford83-90.

Ouknine, L. (2020). Chutes et trouvailles de météorites en Afrique. *Etude géologique et facteurs de distribution et d'altération* Thèse de l'Université Ibn Zohr.

Ouknine, L., Ibhi, A., Khiri, F., Ait Touchn, A., Manzari, P., De Pascale, O., Senesi, G.S. (2019). *Orthopyroxene Megacrysts in Toufassour Mesosiderite: Mineralogy and Origin. EPSC Abstracts 13, EPSC-DPS2019-476.*. EPSC-DPS Joint Meeting.

Ouknine, L., Khiri, F., Ibhi, A. (2015). Les météorites africaines appelées "NorthWestAfrica", patrimoine sans filiation. *22ème Colloque International des Bassins Sédimentaires Marocains.*18-20 DécembreFès, Maroc.

Ouknine, L., Khiri, F., Ibhi, A. (2018). Meteorites classified "Northwest Africa": A dissipated heritage. *Int. J. Innov. Appl. Stud., 24*(3), 1166-1177.

Rasmussen, K.L. Quarterly journal of the royal astronomical society. (1990). *31*, 95-108.

Rickman, H., Błęcka, M.I., Gurgurewicz, J., Jørgensen, U.G., Słaby, E., Szutowicz, S., Zalewska, N. (2019). Water in the history of Mars: An assessment. *Planet. Space Sci., 166*, 70-89. [http://dx.doi.org/10.1016/j.pss.2018.08.003]

Rowe, M.W., Steelman, K.L. (2003). Comment on "some evidence of a date of first humans to arrive in Brazil". *J. Archaeol. Sci., 30*(10), 1349-1351. [http://dx.doi.org/10.1016/S0305-4403(03)00021-9]

Russell, S.S., Zolensky, M., Righter, K., Folco, L., Jones, R., Connolly, H.C., Jr, Grady, M.M., Grossman, J.N. (2005). The meteoritical bulletin 89, September. *Meteorit. Planet. Sci., 40*(S9), A201-A263. [http://dx.doi.org/10.1111/j.1945-5100.2005.tb00425.x]

Russell, S.S., Zolensky, M., Righter, K., Folco, L., Jones, R., Connolly, H.C., Jr, Grady, M.M., Grossman, J.N. (2005). The Meteoritical Bulletin, No. 89, 2005 September. *Meteorit. Planet. Sci., 40*(S9), A201-A263. [http://dx.doi.org/10.1111/j.1945-5100.2005.tb00425.x]

Ruzicka, A., Grossman, J., Bouvier, A., Agee, C.B. (2017). The Meteoritical Bulletin, No. 103. *Meteorit. Planet. Sci., 52*(5), 1014. [http://dx.doi.org/10.1111/maps.12888]

Ruzicka, A., Grossman, J., Bouvier, A., Herd, C.D.K., Agee, C.B. (2015). The Meteoritical Bulletin 101. *Meteorit. Planet. Sci., 50*(9), 1-136.

Ruzicka, A., Grossman, J., Bouvier, A., Herd, C.D.K., Agee, C.B. (2015). The Meteoritical Bulletin, No. 102. *Meteorit. Planet. Sci., 50*(9), 1662-1662.

[http://dx.doi.org/10.1111/maps.12491]

Sadilenko, D.A., Lorenz, C.A., Ivanova, M.A., Roshina, I.A., Korochantsev, A.V. (2013). A small impact crater in the High Atlas, in the Agoudal iron strewnfield. *76th Annual Meeting of the Meteoritical Society.*EdmontonJuly/August 5215, 1.

Sagan, C., Druyan, A. (1986). Cometa EditoraGradiva.(pp. 25-45). Lisboa: 25-45.

Schulz, T., Povinec, P.P., Ferrière, L., Jull, A.J.T., Kováčik, A., Sýkora, I., Tusch, J., Münker, C., Topa, D., Koeberl, C. (2020). The history of the Tissint meteorite, from its crystallization on Mars to its exposure in space: New geochemical, isotopic, and cosmogenic nuclide data. *Meteorit. Planet. Sci., 55*(2), 294-311. [http://dx.doi.org/10.1111/maps.13435]

Singmaster, H. Water-bearing salt crystals come from dawn of solar system. Available at: www.spaceref.com/news/viewpr.html?pid=1989(2000).

Tong, C.H., Lana, C., Marangoni, Y.R., Elis, V.R. (2010). Geoelectric evidence for centripetal resurge of impact melt and breccias over central uplift of Araguainha impact structure. *Geology, 38*(1), 91-94. [http://dx.doi.org/10.1130/G30459.1]

Wasson, J.T. (2011). Relationship between iron-meteorite composition and size: Compositional distribution of irons from North Africa. *Geochim. Cosmochim. Acta, 75*(7), 1757-1772. [http://dx.doi.org/10.1016/j.gca.2010.12.017]

Weber, I., Morlok, A., Bischoff, A., Hiesinger, H., Ward, D., Joy, K.H., Crowther, S.A., Jastrzebski, N.D., Gilmour, J.D., Clay, P.L., Wogelius, R.A., Greenwood, R.C., Franchi, I.A., Münker, C. (2016). Cosmochemical and spectroscopic properties of Northwest Africa 7325-A consortium study. *Meteorit. Planet. Sci., 51*(1), 3-30. [http://dx.doi.org/10.1111/maps.12586]

Weider, S.Z., Nittler, L.R., Starr, R.D., McCoy, T.J., Stockstill-Cahill, K.R., Byrne, P.K., Denevi, B.W., Head, J.W., Solomon, S.C. (2012). Chemical heterogeneity on Mercury's surface revealed by the MESSENGER X-Ray Spectrometer. *J. Geophys. Res., 117*(E12), n/a. [http://dx.doi.org/10.1029/2012JE004153]

Weisberg, M.K., Smith, C., Benedix, G., Folco, L., Righter, K., Zipfel, J., Yamaguchi, A., Aoudjehane, H.C.H.E.N.N.A.O.U.I. (2009). The Meteoritical Bulletin, No. 95. *Meteorit. Planet. Sci., 44*(3), 429-462. [http://dx.doi.org/10.1111/j.1945-5100.2009.tb00742.x]

Woodhouse, B. (1986). Bushman paintings of comets? *Monthly Notes from the Astronomical Society of South Africa, 45*, 33-35.

CHAPTER 6

Meteorites of Sahara

José García[1,*], Lahcen Ouknine[2,3] and Giorgio S. Senesi[4]

[1] *Petrographic Laboratory in Canary Museum of Meteorites, Tanta, Las Palmas, Canary Islands, Spain*

[2] *Geoheritage and Geomaterials Laboratory, Ibn Zohr University, Agadir, Morocco*

[3] *University Museum of Meteorites, Ibn Zohr University, Agadir, Morocco*

[4] *CNR, Institute for Plasma Science and Technology (ISTP), Bari seat, 70126 Bari, Italy*

Abstract: The number of meteorites collected in the North African Sahara Desert is very relevant, as their conservation is facilitated by the dry climate, thus it represents one of the most important regions to recover rocks from space, along with Antarctica, Atacama and the great deserts of North America. However, more than 90% of the desert surface feasibly preserving important meteorites is estimated to be not yet explored. New finds are classified annually, and countries such as Mauritania, Mali and Egypt are emerging. The most optimistic forecasts suggest that many new meteorites will continue to be extracted from the great Sahara Desert in the coming decades. Collecting and preserving them properly is essential to bequeath to science such important research materials such as meteorites.

Keywords: Climatic conditions, Documentation, Meteorite finds, Preservation, Searching meteorites, Sahara Desert.

INTRODUCTION

The Sahara Desert is claimed to cover more than 9.400.000 km², (Española, 2004), thus being the largest hot desert in the world and the third largest desert after the frozen deserts of the Arctic and Antarctica. The Sahara Desert stretches majestically through Algeria, Chad, Egypt, Libya, Mauritania, Mali, Morocco, Tunisia, Sudan and Niger, bordering at east with the Red Sea, at west by the Atlantic Ocean, at north with the Mediterranean Sea, and at south with the Sahel, from which Sub-Saharan Africa begins. This vast extension of land is dominated by "erg" or dune fields shaped by the wind, and by "Hammada," or extensive valleys of rocky soils (Grove, 1958). Furthermore, the Sahara includes plains of gravel soils, called "reg", huge dry valleys and salt flats are known as "Shatt".

---

[*] **Corresponding author José García:** Petrographic Laboratory in Canary Museum of Meteorites, Tanta, Las Palmas, Canary Islands, Spain; E-mail: direccion@museocanariodemeteoritos.com

In these primeval lands of the great Sahara Desert, a treasure is hidden (Milich, 1997).

In April 2015, the senior author conducted his first meteorite search trip to the Sahara with the aim of collecting samples for laboratory studies. Its plane landed at Laayoune, from where he travelled to different regions of the Sahara on a trip of several days, during which he met numerous meteorite seekers. Meteorites recovered in Sahara are not only abundant in number, but vary in typologies, including the rarest and most unknown rocks in the solar system. In Sahara, the search for meteorites does not obey any rule, and it is done legally, but in a disorderly manner. The senior author frequently met people living in their Khaimas in the desert keeping black stones that, after an examination, turned out to be meteorites. It is also frequent to stand on the terrace of a bar or cafe and talk about meteorites, and have someone show up with meteorites in an attempt to sell them to the highest bidder. Despite the economic and social situation in the Sahara is quite precarious, and that some people still try to get money selling meteorites, the situation has changed a lot in recent decades, as meteorite typologies have been established, their characteristics, rarity and abundance are known, and their cost is established.

The history of meteorites in the Sahara dates back to 1986-1987, when a German team installing seismic stations for the exploration of oil beds discovered 65 meteorites in a desert plain about 100 km southeast of Dirj, Libya. This was the first indication that large numbers of meteorites could be found in Sahara. A few years later, an anonymous engineer who was a fan of the desert looked at some photographs of meteorites found in Antarctica and recalled observing similar rocks in areas he had travelled in North Africa. Thus, in 1989 he returned to Algeria and recovered about 100 meteorites from at least 5 locations. In the next 4 years, he and other followers found at least 400 more meteorites in the same locations, and in some new areas in Algeria and Libya. The locations where they found the meteorites are known as reg and hamada, which are flat areas covered only by pebbles and small amounts of sand, where they have been very well preserved due to the arid climate of the region and can be easily spotted (McCall *et al.*, 2006). Although meteorites have been sold commercially and collected by hobbyists for several decades until the time of the Sahara finds in the late 1980s and early 1990s, most meteorites were deposited or purchased by museums and similar institutions where they were exhibited and made available for scientific research. However, the rapid availability of a large number of meteorites that could be found relatively easily in places that were easily accessible, led to the rapid increase in the commercial collection of meteorites, which was accelerated in 1997 when the first meteorites from the Moon and Mars were found in Libya. By the end of the 1990s, private meteorite search expeditions had been launched across the Sahara, so that, although some specimens were recovered by researchers, most of the material was sold to private collectors. By now, these expeditions have recovered about 14000 officially classified meteorites in Northwest Africa, and hundreds of thousands never officially classified, so generating a lack of information on both the number and type of meteorites.

The official database of the Meteoritical Society, *i.e.*, the Meteoritical Bulletin, includes more than 13100 meteorites with the NWA designation, of which 109 are Acapulcoites-Lodranites, 30 Angrites, 21 Aubrites, 42 Brachinites, 1199 carbonaceous chondrites, 126 are from the enstatite group of chondrites, 1549 belong to some types of Vesta HED family, 129 are iron, 374 originate from the Moon, 250 from Mars and belong to the group SNC, Mesosiderites are 146 and Pallasites 51, Ureilites are 357, Winonaites 48, Rumurutites 198, the group of ordinary chondrites has 6386 classified specimens. The group of ungrouped meteorites counts up to 20 irons, 39 chondrites and up to 96 achondrites. Furthermore, more than 2000 NWA meteorites are registered in the official database with a provisional status, the analysis and typology of which have never been updated. Thus, practically all known types of meteorites have been recovered in the Sahara. Actually, Sahara meteorites are very often referred with "provisional" names in the databases with not updated or erroneous information, or as "pairing" to others. Sometimes the same meteorite has been classified at different times with different names, which is a problem that needs to be addressed by the authorities who control the classification processes. Since the first meteorite markets were created, especially in Morocco, supported by nomads and local people who dug into the desert in search of specimens for sell, tons of meteorites, the so-called NWA meteorites have been distributed, most of which with no information on how, when and where they were discovered.

A lucrative trade was established in some localities, especially Erfoud, Marrakesh and Zagora, and more recently Laayoune and Agadir, and even in Nouakchott, the capital of Mauritania, where no law regulates the searching and selling of meteorites. Differently, in Algeria, anything that has to do with the ground (sand, fossils, minerals, meteorites, *etc.*) is controlled by severe restrictive measures, and punished with fines and even jail. In Morocco, the possession and trade of meteorites are legal, but their export abroad is subject to regulatory restrictions. Since June 2019, the Moroccan mining regulation includes meteorites as a mining resource that subject them to some kind of exploitation law. In particular, Article 3 of the mines code establishes that the exploitation of mine resources can only be exploited by authorization of the state. The exploration, investigation and exploitation of mine products can be carried out by virtue of a mining title issued in accordance with the provisions of this law and related text. The search permit and mines license constitute real estate rights, of limited duration and distinct from land ownership, that is, even the owner of the land needs an authorization to exploit meteorites. However, the question is: if scientists do receive a large donation of research samples found in the desert, will such regulation limit the flow of scientific work and resources? Can a country that restricts media to science really be considered a developed country?

In recent years, multiple factors have affected the commercial exploitation of Sahara meteorites. The first and most important factor is the continuous restriction to which police authorities, based on the free interpretation of some laws, subject search instruments in many countries. The second factor is related to the fact that modern scientific means have allowed to establish in detail the groups of known meteorites and know the amount of their recovery, thus the demand for meteorites, no matter how much it grows, will always have an available supply, because there are tons of annual recoveries. The higher the offer, the lower the price.

If exorbitant amounts of money were paid decades ago for a meteorite fragment, nowadays, prices have collapsed and dropped considerably, as anyone can have meteorites for a handful of coins. Nobody can sell meteorites at high prices anymore, because nobody buys meteorites at those prices. However, some types of few and rare meteorites still have considerable value, even if their finding and proper identification are difficult, thus, the support of scientists is essential.

Finding meteorites is easy for many seekers, especially if they explore virgin regions, but only a few of them will make a living from the meteorite business because the most valuable ones are the rarest and the most difficult to recognize. Often, the recognition of valuable meteorites depends exclusively on a simple petrographic study. However, Sahara meteorites often have been subject to modifications and alterations due to the environmental conditions of the desert, thus it is very important to recover them as soon as possible and in conditions adequate for their documentation and conservation.

Searching for meteorites in the desert can be a desperate activity if the seeker is not provided with a minimum of knowledge, so he will often fill his backpack with black stones of no value. The choice of the search locality is also very important, thus it is always advisable to carry a geolocation device (GPS or mobile with application) to document the discovery point, also to establish possible new dense areas, and in the best case, classify the meteorite with its own name. In the fortunate case, one can experience a fall and trace the trajectory of the meteor to its location on the ground, as it happened with the Bensour, Tarda and Oudiyat Sbaa meteorites, and many others. Generally, fireballs appear quite high in the sky, sometimes more than 100 km in altitude, whereas they are visible only at most up to 20 km in altitude. Furthermore, depending on the angle, the fragments can fall near where they were seen exploding, or tens of km away. The scattering fields can also be very large, as in the case of Al Haggounia 001, which extends up to 40 km².

Searching for meteorites in the desert is recommended to be carried out in optimal places such as reg, hamada and rocky land, where rain is scarce, and no dunes or sandy soils are present. Places covered by few small stones (reg) are ideal for meteorite search, whereas their recognition is more difficult in a terrain covered by large rocks and in wadis that are full of basaltic, very fragmented black rocks originated from mountains that can be confused with real meteorites. Differently, dry riverbeds and valley troughs represent good places as torrential rains tend to wash rocks down to lower areas so that often meteorite clusters can be found. Something similar occurs in Antarctica in the valleys of glaciers, where meteorites are dragged down and deposited. Furthermore, pale-colored soil favors the finding of especially black-colored meteorites (Fig. **1**) even at a considerable distance as they stand out on the ground, whereas dark ground complicates the search.

Finally, when a search for meteorites is organized, a very important action is to document the place of recovery by using a GPS or a mobile phone with the appropriate application, especially if a meteorite cluster is found. When a meteorite reaches a laboratory, scientists will ask many questions about the specimen, including where it was collected. Thus, the information acquired by following a few simple rules is as important as the specimen itself.

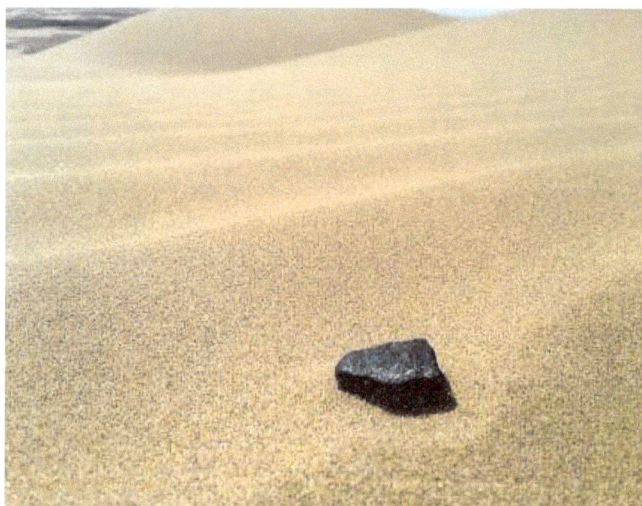

Fig. (1). A Martian meteorite of 5 cm diameter in the site of Nayla. (J.García/LPMCM, 2018).

PROTOCOL TO BE FOLLOWED IN METEORITE RECOVERY

Some simple rules that should be regarded as a practical protocol for recovering meteorites include:

Do Not Touch the Meteorite

A meteorite is a virgin object that originated from another region of the Solar System, thus it must be preserved as much as possible. Although meteorites do not contain radioactive and other components dangerous to humans, the first rule is to avoid any direct contact with the meteorite with hands, as human skin generates fats and other organic components that, when transferred to the meteorite, can react chemically and promote degradation and oxidation processes, that also interfere with its subsequent analysis.

Mark the Coordinates and the Scale

A GPS or a mobile application must be used to record and document the geographical coordinates of the place where the meteorite is found and assign a proper name to the sample. In the case of relevant findings, the exact coordinates will allow us to infer the dispersion field, and contribute to solve unanswered questions on the dispersion of meteorite falls in the world. The use of a compass to calculate the orientation of the object on the ground is also highly advisable. It is also important to measure the sample size using scale cubes or a ruler, a tape measure, a coin or any other appropriate means.

Provide Photographic Documentation of the Finding

Before removing the meteorite from the ground, several pictures of the stone on the ground, the terrain around it, the horizon, *etc.*, should be taken from different angles and at different distances. All the photographic documentation achieved constitutes documentary material of extraordinary value and importance for scientists. It also represents an important graphic document to accompany the meteorite in its future exhibitions, scientific works, *etc.* A well-documented meteorite acquires a greater value than one collected without scientific proof.

Take Notes

When a meteorite is found, as much information as possible, *i.e.*, date, time, name and coordinates of the place, circumstances of the discovery, annotations, sampling procedure, *etc.*, should be duly annotated. The more information, including diagrams, drawings, and plans, are collected, the better will be the contribution provided to scientists and collectors. In any case, the documentary value is unquestionable, and all the time necessary should be devoted to record as many notes as possible.

Collection and Preservation of Fragments

Using gloves, or any means that prevents the direct touching of the specimen, it will be placed inside a sample bag wrapped in aluminum foil. However, in the case the sample is wet, it should be wrapped in absorbent paper and the bag left open in a warm place to allow drying. Inside the bag, a note should be added with a reference indicating that the sample is identified by the annotations and photos taken of it and its finding. One bag should be used for each sample to prevent them from spoiling against each other, and all bags should be kept in a safe place where they will not suffer blows. Although meteorites have been subjected to quite harsh cosmic conditions, may be older than our planet and generally seem hard, during their time of residence on our planet, they have faced important alteration processes that may have made them fragile. Recent falls show a melting crust that consists in the black shell that forms when they burn along their crossing through the atmosphere, which provides them with a very particular beauty highly appreciated by collectors. Preserving these characteristics is very important for meteorites scientific and commercial value. A meteorite that maintains its fusion crust in good condition, without deterioration, will always be valued much more than a meteorite that has been hit with other stones, is broken, worn, oxidized, or shows significant deterioration. The aesthetics of meteorites is an important attribute that directly affects their economic value. Fig. (**2**) shows a recently found meteorite whose fusion crust is very well preserved despite some natural fractures. After their recovery, meteorites should not be touched, wet or exposed to air, in order to keep them in the best conditions for a long time. Fig. (**3**) shows a recently fallen meteorite that has been stored with other stones, has largely lost its scab and has been exposed to air for a long time, so oxides have formed that turned the meteorite matrix brown. Fig. (**4**) shows a meteorite that has retained the entire surface, but was not protected, so that oxidation has turned brown the entire crust.

Fig. (2). An NWA meteorite with a very well-preserved fusion crust. (J.García / LPMCM meteorites collection).

Fig. (3). An NWA meteorite exposed to air for a long time. (J.García / LPMCM meteorites collection).

Fig. (4). An NWA meteorite with an oxidised crust. (J.García / LPMCM meteorites collection).

FACTORS AFFECTING METEORITE PRESERVATION

The beauty and value of well-preserved meteorites compared to those that have not been well-treated leaves no doubt about the importance of knowing how to treat and protect them. Various factors, which are shortly described below, can affect meteorites' status and can completely destroy them.

Wind

This erosion is mechanically produced by a continuous flow of sand particles moved by the wind that collide with meteorites and produce a kind of polishing of their surface called "desert varnish". Meteorites become shiny and smooth, loose their crust, and subsequently mass in a continuous process that will expose them to oxidation. This process cannot be controlled, thus a rapid recovery is vital to avoid meteorite continued erosion that will finally turn them to sand, like all desert rocks.

Water

Both rain and moisture are very aggressive to meteorites. Water attacks meteorite minerals, oxidizing them and producing new molecular forms that will cause instability, deposition of other materials and their transformation into forms stable on the Earth. Thus, meteorites must be quickly dried in the sun, in the heat, and protected with cotton fabrics to avoid any contact with moisture, and they must never be washed. Putting them in sacks or bags if they are wet can be a ruin. The ideal way to store them is in the presence of silica gel to prevent humidity.

Disorganized Collection of Samples

The confused collection of meteorites by stacking them all together in sacks or boxes, with the risk of colliding with each other, suffering friction and fractures, must be absolutely avoided. Meteorites must be individually protected as they are very sensitive materials and must not collide or split, otherwise, they will lose value. In particular, pieces of special scientific interest must be well preserved so that their analysis and classification can be carried out in the best conditions.

Magnets

The use of magnets is quite common for meteorite identification, as 90% or more of them contain iron thus, they can be attracted to a magnet. However, many terrestrial rocks also contain iron and they react to a magnet in a similar way. Noteworthy, meteorites still retain part of the magnetic fields of the place where they were formed, and these magnetic remnants constitute very valuable scientific information, thus their preservation from strong magnetic fields will always be very important for the scientist (Weiss *et al.*, 2010).

WEATHERING

By principle, meteorites preserve the features of the environmental conditions in which they formed, thus they represent an invaluable source of information on the conditions of formation of our planetary system. However, the information kept in meteorites can be seriously affected from the moment they fall onto our planet as Earth is very rich in oxygen that easily reacts with most chemical elements, including iron, producing oxides. Native iron is present in most meteorites in the form of kamacite and taenite alloyed with nickel in different proportions, commonly 5.5% in kamacite, and a higher amount in taenite, thus oxygen reacts easily with these metals producing oxides. Thus, the information contained in them is progressively lost until it disappears. The knowledge of the extent to which terrestrial conditions affect meteorites is very important to infer questions such as rescuing the information that still remains in them, and their time of exposure to outer space.

The degree of degradation, generally called weathering (W) degree is established at the moment in which the meteorite is found; this degree will stabilize or change to a higher or lower value, on the dependence on its preservation conditions (Fig. 5). The weathering scale was proposed by Jull *et al.* (1991) and updated by Wlotzka (1993), Wlotzka *et al.* (1995) and Al-Kathiri *et al.* (2005). On this scale, six degrees are set for hot desert meteorites: (i) the W0 grade applies to freshly fallen and collected, apparently not altered meteorites featuring non-oxidized metals and a clean matrix; (ii) grades from W1 to W4 indicate increasing levels of oxidation of metals and troilite (FeS); (iii) grade W5 indicates a low alteration of mafic silicates, and grade W6 indicates their extended alteration. The weathering grade should be assigned precisely by examining thin sections of the meteorite, but it can be approximated even with the naked eye on a cut meteorite surface.

Fig. (5). Different weathering degrees of chondrites. (J.García / LPMCM).

Due to the oxidizing conditions existing on the planet, meteorites gradually oxidize, passing from a state of freshness characterized by a well-preserved not-oxidized black fusion crust to a more oxidized state, brown to ochre in color, and sometimes highly polished. Due to wind and desert sand, rust spots produced by the degradation of the metallic phase appear on the meteorite surface, with impact cracks often filled with grains of sand and sandy sediments. The next step will bring a totally unrecognizable meteorite, with metals totally oxidized and replaced and the melting crust no longer present or deeply deteriorated. The specimen appears extensively eroded and in an extremely poor state of preservation that will finally result in complete integration into the terrestrial ground. Magmatic differentiation with progressing time will complete the process.

In particular, with increasing metal oxidation, a brown color appears first around metallic grains (Figs. **6** and **7**), then affecting nearby silicates and the matrix and finally producing their massive alteration (Figs. **8** and **9**). Extensively weathered meteorites often become friable and very brittle. In terms of alteration degree, two types of oxides have been distinguished in meteorites, one is the oxides of Fe-Ni-S metals (kamacite, taenite, troilite), and the other is sialic oxides of Fe, Si and Al formed in the degradation of mafic silicates (Gooding, 1986). The same author found that troilite is less susceptible to oxidation than the native metal, taenite is less oxidisable than kamacite, and silicates are less oxidisable than metals and sulfides.

The degradation process of meteorites involves not only the formation of oxides, but also the hydrolysis, hydration, carbonation and dissolution of silicates. The formation of clay minerals has been documented in quite a few meteorites, such as Al Haggounia 001 and paritarians. An important aspect of this process consists in the formation of carbonates and evaporite salts that are introduced as interstitial veins and drastically alter the composition of the meteorite.

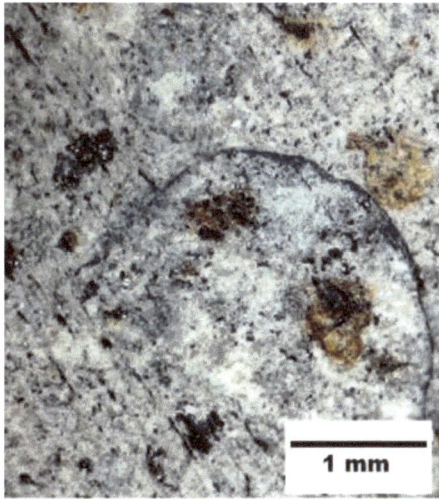

Fig. (6). The grade W1 weathered Bensour meteorite. Rust spots are visible around some metal particles. (J.García / LPMCM).

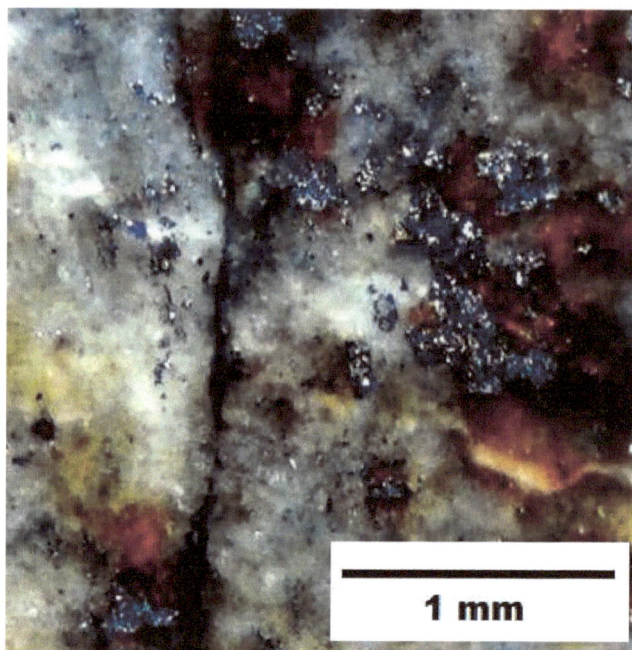

Fig. (7). A grade W2 weathered not classified NWA meteorite showing a wider oxide alteration around the metallic phases. (J.García / LPMCM).

Fig. (8). The grade W3 weathered Gao Guenie meteorite. The oxidation of the metallic phase and the extension of oxides to silicates are greater. The specimen shows a reddish hue due to this alteration. (J.García / LPMCM).

Fig. (9). The grade W4 weathered NWA 4686. The metallic phase appears almost 95% oxidized, whereas mafic silicates start to show alterations. (J.García / LPMCM).

RECOGNITION OF METEORITES ON THE GROUND

Thanks to the pale, golden, ochre floor of the Sahara Desert, black stones stand out on it, thus very often, meteorites can be easily recognized on the ground not only by trained eyes, but by anyone. However, the desert also contains several black rocks, *e.g.*, fragments of basalt of igneous origin, which makes the search more complicated. Furthermore, remains of metals, shrapnel, tires, mine wastes and a thousand more junk from past wars may occur in the desert, which can be rusted and look like meteorites.

Even prehistoric lithic tools, flint arrowheads and stones that have been clearly manipulated and worked by humans can be found quite frequently. These archaeological pieces, which should not be confused with meteorites nor should be unprotected and damaged, are made of basalt, flint and other hard rocks (Fig. **10**). Furthermore, flints and basalts that show the characteristic "desert varnish" really confuse the seekers in their search of meteorites. As mentioned above, the varnishing effect is caused by the mechanical action of the sand transported by the wind that hits the object surface and literally polishes them. Flints are an amorphous variety of quartz, so iron is not present in their composition, thus they do not react to magnets. They display an incredible variety of colors, *i.e.*, red, brown, white, ochre and different shades that can be mistaken for meteorites even to the eyes of an expert, although a careful examination will rule them out (Fig. 11).

Fig. (10). A lithic axe. (Anonimous. Milanuncios.es).

Fig. (11). A 10-cm diameter desert varnished flint photographed *in situ* (from https://en.wikipedia.org/).

DENSE METEORITE COLLECTION AREAS

Although finding a real meteorite is not easy, it happens. It is estimated that, on average, a piece can be found in about 500 km² of land or less in unexplored virgin areas, or on stable terrain and dense areas.

A dense meteorite collection area has defined a place where falls accumulate and numerous meteorites can be recovered in a specific space on the ground. Always important is to respect the physiognomy of the places, avoiding to leave garbage, modify elements of the terrain, such as rocks and anything else typical of the place. The Sahara Desert features many dense meteorite collection areas where significant finds have been recorded (Fig. **12**).

Fig. (12). Dense meteorite collection areas in the Sahara. Data retrieved from The Meteoritical Bulletin (https://www.lpi.usra.edu/meteor/metbull.php) and reported on Google Map (https://www.google.es/maps).

SCIENTIFIC AND MARKETING VALUE OF METEORITES

The laws of each country regulating the search, collection, trade and export of meteorites should be well known by searchers, as some countries prohibit it with serious penalties. As mentioned above, meteorites had a great economic value at the time when the first specimens were discovered, with collectors and scientists paying incredible sums to get them. However, as time passed, the knowledge about these rocks was refined and different categories or groups were established, which contributed to the diversification of the price. If decades ago, a gram of lunar meteorite could cost up to $ 40,000, today, it is affordable to anyone. During his last trip to the Sahara, the senior author could buy a significant quantity of lunar meteorites at a fairly affordable price.

The quantity and rarity of the specimens on offer, but also their aesthetics and other characteristics, contribute to their price. In the case of ordinary chondrites, which account for up to 87% of falls and finds, the price can be as low as a few cents per gram. But the price increases if the specimen is well preserved, aesthetically beautiful, or of petrographic type 3, which is the purest type of asteroid materials and the most interesting of this group. The price also differs between meteorite falls, finds, with the formers having a higher economic value than the latter within the same type. Another important factor that contributes to the value of a meteorite is its type, and rarity.

In particular, achondrites of planetary origin, such as those coming from Mars and the Moon, are meteorites of great interest to science, as they represent engraved records of their geological history, and can provide invaluable information on their formation, evolution over time and their current conditions. Erosion by wind attacks these meteorites very violently, as they are mainly composed of soft minerals, feldspars and plagioclase.

A Martian meteorite find will always be more valuable than an ordinary chondrite fall. Actually, ordinary chondrites are the most common to recover, whereas groups of meteorites, such as the SNC from Mars and the LUNs from the Moon, represent authentic rarities. Rocks of the crust and mantle of Mars and the lunar surface are real treasures for scientists due to the invaluable information they contribute to the knowledge of extraterrestrial planetary geology.

Although they are more abundant than the previously mentioned ones, very rare and precious are also the HED meteorites from the asteroid Vesta, which is one of the largest asteroids in the main belt, 520-km of average diameter, located at about 150-million-km from the Earth orbit, a figure that triples if both bodies are opposite in their respective orbits. In particular, Eucrites, also called white lavas, are sensitive and delicate materials consisting of basalts solidified from magmatic flows, which are composed of igneous material rich in pyroxene and plagioclase, such as genomícticas or monomictic eucrites (Fig. **13**). Diogenites are crystals of orthopyroxene and olivine crystallized from the precipitate of magmatic chambers, the largest known falls of which was classified as NWA 783, with the official recognition of fragments weighing more than 80 kg and even today fragments are still found in Saquia El Hamra; and Howardites, which cement different types of clasts and materials, such as regolithic sedimentary deposits, reveal the troubled geological history of Vesta. Planetary achondrites are very rare materials, so much so that collectors cherish them with great interest, and pursue them for incorporation into their collections and businesses. In any case, the rarity of the piece, constitutes one of the main factors to measure its scientific and economic value.

Fig. (13). Texture of NWA 10909, a monomict eucrite discovered in the Sahara in 2015. (J.García / LPMCM).

In the meteorite business, there is a lot of competition, *i.e.*, one can find somebody who sells eucrites at a price, and others who sell them for half or less, thus, meteorites from a commercial perspective are also subject to the law of the market like any other product. The low price of a specimen is also related to the fact that it needs to be analyzed and classified, with loss of its mass in the investigation, so the price of analysis, which sometimes is very high, must be added. The classification process is also very important, as certifying the authenticity of a meteorite is essential, but costly. Thus, many seekers and collectors chose not to classify their findings, as the cost of analyses far exceeds the commercial value of the meteorite, so they are left as unclassified meteorites. There is no point in investing 1000 dollars in classifying a meteorite whose commercial value is no more than 100. However, in many cases, rocks that appear to be meteorites are just simple terrestrial rocks, thus, it is recommended that at least the authenticity of a specimen be certified. All not classified specimens are included in a common group under the designation of Northwest Africa (NWA), not numbered meteorites that are available at a fairly affordable price.

Seekers and collectors usually donate fragments to museums and laboratories, which allows them to study and learn more about meteorites. Obviously, all samples donated for scientific research cease to have a commercial value and cannot be sold, as they will be used solely and exclusively for scientific use. For analysis and classification processes, it is indeed necessary to donate meteorite pieces according to the established protocols. When somebody donates a meteorite, his/her name will appear in scientific publications as a donor, so the credibility of the person in the sector will be consolidated, and the interest in his/her findings will grow.

A FEW FINAL RECOMMENDATIONS

An important recommendation is to never run alone in the desert in search of meteorites but accompanied by someone who knows the area and be a guide to safe places with the support of a GPS, a geolocator that allows you to know at all times exactly where you are, and to indicate the way to move safely in the desert. When crossing the desert, it is necessary to be well-equipped with food and water, appropriate clothes and tents. At night, knowing the constellations and position of stars can be very useful to orient in the desert, so the feeling of insecurity and anxiety will disappear. The simple fact of recognizing the geographic north makes the person feel a certain security. The desert is full of topographical references that may be very useful for orientation. Any relief, any natural or artificial element, a mountain in sight, an electric poster, a pile of stones or a pair of tires placed on purpose also can act as very useful topographic landmarks. Another important recommendation is that search trips in the desert should always be documented, *e.g.*, by carrying a GoPro camera installed in clothing, backpack or vehicle.

All scientific and optical instruments must be carried well-protected in closed cases or briefcases, wrapped in bags or covers, as sand can penetrate anywhere on the equipment. Frequently cameras are damaged by sand and optical lenses are scratched by grains of sand, and so on. As well, the protection of meteorite finds is very important, and the specimens should be separately well-wrapped in aluminum or paper and carried in appropriate bags and containers.

CONCLUSION

The Sahara desert is full of objects that appear to be meteorites but are not, thus, only an expert can conclude if a rock is a meteorite or not. The climatic conditions of the desert affect all its materials, terrestrial and non-terrestrial, which are finally integrated into the environment, although in a very different way than they would be in other parts of Earth. Meteorites fallen in tropical or humid areas are affected more violently, and their degradation is rapid, while those fallen in desert areas may take centuries to degrade. In particular, meteorites fallen in Antarctica and polar areas will freeze and may remain unaltered for millions of years, which makes them pieces of special scientific interest, as their better degree of conservation implies that the pre-falling conditions remain, which allows scientists to infer questions about the conditions of their formation and development prior to their falling. The residence time, *i.e.*, how long a meteorite has been exposed on the Earth, is a key factor, thus the recovery of meteorites as quickly as possible after their fall guarantees their preservation and scientific quality, besides having a positive influence also on their value.

From the moment that a meteorite leaves the desert, it will be subject to different environmental conditions affecting its alteration state. In particular, in humid or oxidizing conditions, the alteration processes will be accelerated, while if properly preserved, the progress of degradation can even be stopped. Meteorite searchers often go to regions where major falls/finds were recorded in the past, and even today, they still find fragments in these scattering fields. However, apparently, the first and last recoveries seem not to be the same, *i.e.*, the physical aspect and sometimes some chemical measurements are different, which is due to the alteration that the rocks undergo over the years due to changes in temperature, humidity, erosion, *etc.*

A seeker can travel hundreds of km in the desert and spend long hours, weeks, months, even years, until he can recover a meteorite after a long, tedious, endless search. Meteorites are rare and prized targets for seekers who take their hunting as a real job. Although their hard and timeless work can be rewarded, the meteorite searchers have to invest a lot of time and resources in the deserts. However, when a meteorite searcher becomes allied with science by collaborating with the donation of samples for research, he/she will gain a greater reputation by not losing anything with a donation, but investing in a crucial factor, *i.e.*, credibility.

To study meteorites, they must be cut to be able to analyze their interior, but the searchers generally do not want to cut their specimens, as they believe that this will decrease their value. The international protocols, however, establish that a certain mass of each meteorite sample should be donated for analysis and its official classification. Thus, cutting a meteorite does not decrease its value, but makes it easier to sell it than a whole piece of dubious nature as many collectors prefer to buy a cut meteorite that has been verified inside and authenticated by an official Institution, also to enjoy its internal structure.

In conclusion, the Sahara desert, together with Antarctica, Atacama and the great deserts of North America, has become one of the most important places to recover meteorites. Nevertheless, more than 90% of the Sahara surface is estimated to be not explored, and to still contain important meteorites. New finds are indeed classified annually, and countries such as Mauritania, Mali and Egypt are emerging in this dizzying competition. The most optimistic forecasts suggest that many new meteorites will continue to be extracted from the great Sahara Desert in the coming decades. The proper preservation of which will be instrumental for meteorite science.

REFERENCES

Afiattalab, F., Wasson, J.T. (1980). Composition of the metal phases in ordinary chondrites: implications regarding classification and metamorphism. *Geochim. Cosmochim. Acta, 44*(3), 431-446.
[http://dx.doi.org/10.1016/0016-7037(80)90042-3]

Anders, E. (1964). Origin, age, and composition of meteorites. *Space Sci. Rev., 3*(5-6), 583-714.
[http://dx.doi.org/10.1007/BF00177954]

Antoniadi, E.M. (1939). On ancient meteorites, and the origin of the crescent and star emblem (with plates VII, VIII). *J. R. Astron. Soc. Can., 33*, 177.

Bevan, A.W.R., Binns, R.A. (1989). Meteorites from the nullarbor region, western australia: i. a review of past recoveries and a procedure for naming new finds. *Meteoritics, 24*(3), 127-133.
[http://dx.doi.org/10.1111/j.1945-5100.1989.tb00954.x]

Bischoff, A. (2001). Meteorite classification and the definition of new chondrite classes as a result of successful meteorite search in hot and cold deserts. *Planet. Space Sci., 49*(8), 769-776.
[http://dx.doi.org/10.1016/S0032-0633(01)00026-5]

Bischoff, A., Geiger, T. (1995). Meteorites from the Sahara: Find locations, shock classification, degree of weathering and pairing. *Meteoritics, 30*(1), 113-122.
[http://dx.doi.org/10.1111/j.1945-5100.1995.tb01219.x]

Bland, P.A., Artemieva, N.A. (2006). The rate of small impacts on Earth. *Meteorit. Planet. Sci., 41*(4), 607-631.
[http://dx.doi.org/10.1111/j.1945-5100.2006.tb00485.x]

Claussen, M., Kubatzki, C., Brovkin, V., Ganopolski, A., Hoelzmann, P., Pachur, H.J. (1999). Simulation of an abrupt change in Saharan vegetation in the Mid-Holocene. *Geophys. Res. Lett., 26*(14), 2037-2040.
[http://dx.doi.org/10.1029/1999GL900494]

Campbell-Brown, M.D., Hildebrand, A. (2004). A New Analysis of Fireball Data from the Meteorite Observation and Recovery Project (MORP) *Modern Meteor Science An Interdisciplinary View..* Springer. 95(1), pp. 489-499.

Española, R.A. Sáhara o Sahara». Diccionario panhispánico de dudas. (2004). Real Academia Española.(Vol. 5). Santiago: (10).

Grove, A.T. (1958). The ancient erg of Hausaland, and similar formations on the south side of the Sahara. *Geogr. J., 124*(4), 528-533.
[http://dx.doi.org/10.2307/1790942]

Harper, D. Sahara: Online Etymology Dictionary. Available at: https://www.etymonline.com(2021).

Harvey, R. (2003). The origin and significance of Antarctic meteorites. *Chem. Erde, 63*(2), 93-147.
[http://dx.doi.org/10.1078/0009-2819-00031]

Histoire de l'Académie royale des sciences. (1772). Académie royale des sciences..(p. 20). Paris:.

Huss, G. R., Rubin, A. E., Grossman, J. N. (2006). Thermal metamorphism in chondrites. *Meteorites and the early solar system II., 943*, 567-586.
[http://dx.doi.org/10.2307/j.ctv1v7zdmm.34]

Jarosewich, E. (1990). Chemical analyses of meteorites: A compilation of stony and iron meteorite analyses. *Meteoritics, 25*(4), 323-337.
[http://dx.doi.org/10.1111/j.1945-5100.1990.tb00717.x]

La Paz, L. Topics in meteoritics: hunting meteorites: their recovery, use, and abuse from paleolithic to present. (1969). University of New Mexico Press.

Maier, W.D., Andreoli, M.A.G., McDonald, I., Higgins, M.D., Boyce, A.J., Shukolyukov, A., Lugmair, G.W., Ashwal, L.D., Gräser, P., Ripley, E.M., Hart, R.J. (2006). Discovery of a 25-cm asteroid clast in the giant Morokweng impact crater, South Africa. *Nature, 441*(7090), 203-206.
[http://dx.doi.org/10.1038/nature04751] [PMID: 16688173]

Marvin, U.B. (1996). Ernst Florens Friedrich Chladni (1756-1827) and the origins of modern meteorite research. *Meteorit. Planet. Sci., 31*(5), 545-588.
[http://dx.doi.org/10.1111/j.1945-5100.1996.tb02031.x]

McCall, G.J.H., Bowden, A.J., Howarth, R.J. (2006). The history of meteoritics and key meteorite collections: fireballs, falls and finds. Geological Society of London *256*, 167.
[http://dx.doi.org/10.1144/GSL.SP.2006.256]

Meteoritical Bulletin Search the Database Available at: https://www.lpi.usra.edu/meteor/MB-Archive.php

Patterson, C. (1956). Age of meteorites and the earth. *Geochim. Cosmochim. Acta, 10*(4), 230-237.
[http://dx.doi.org/10.1016/0016-7037(56)90036-9]

Sears, P.M. (1965). Notes on the beginnings of modern meteoritics. *Meteoritics, 2*(4), 293-299.
[http://dx.doi.org/10.1111/j.1945-5100.1965.tb01436.x]

Sears, D.W. (1978). The nature and origin of meteorites. New York: Oxford University Press.

Sears, D.W. (2004). The origin of chondrules and chondrites. Cambridge University Press.
[http://dx.doi.org/10.1017/CBO9780511536137]

Smith, C., Russell, S., Benedix, G. (2009). *Meteorites.* (pp. 11-15). London: Natural History Museum.

Strahler, A., Arthur, N., Alan, H. S. (1987). *Modern Physical Geography.* (Third Edition., p. 347). Nueva York: John Wiley & Sons..

Tucker, C.J., Dregne, H.E., Newcomb, W.W. (1991). Expansion and contraction of the sahara desert from 1980 to 1990. *Science, 253*(5017), 299-300.
[http://dx.doi.org/10.1126/science.253.5017.299] [PMID: 17794695]

Walton, K. (2017). *The arid zones..* Routledge.
[http://dx.doi.org/10.4324/9781315131009]

Weiss, B.P., Gattacceca, J., Stanley, S., Rochette, P., Christensen, U.R. (2010). Paleomagnetic records of meteorites and early planetesimal differentiation. *Space Sci. Rev., 152*(1-4), 341-390.
[http://dx.doi.org/10.1007/s11214-009-9580-z]

CHAPTER 7

Meteorite Falls, Finds and Impact Craters in North East Africa: Egypt, Sudan and Libya

Mohamed Th. S. Heikal[1,*], Lahcen Ouknine[2,3] and Giorgio S. Senesi[4]

[1] *Geology Department, Faculty of Science, Tanta University, Tanta, Egypt*

[2] *Geoheritage and Geomaterials Laboratory, Ibn Zohr University, Agadir, Morocco*

[3] *University Museum of Meteorites, Ibn Zohr University, Agadir, Morocco*

[4] *CNR, Institute of Plasma Science and Technology (ISTP), Bari seat, 70126 Bari, Italy*

Abstract: The aim of this chapter is to provide an updated review of the available data concerning the confirmed and proposed meteorite falls and finds and their impact craters in Egypt, Sudan, and Libya. Among the 190 confirmed impact/sites on the Earth crust, only less than 8 have been identified in northeast Africa, in particular, BP and Oasis in Libya and Kamil in Egypt. Very few other structures of alleged impact are located in these countries have been described in the literature. The record of meteorites and their impact craters in North East Africa are still incomplete and debated. The only criteria that provide evidence of an impact are the occurrence of circular geological features and their shock-metamorphic effects on the target rocks (shatter cones and diagnostic shock-metamorphic mineral deformations). These effects are evident in the Kamil crater (Egypt) and most impact craters in Libya. The discovery of preserved meteorite fragments and the detection of the geochemical signature of meteoritic indicators should be considered carefully during fieldwork and advanced further studies.

Keywords: Impact craters, Meteorite falls and finds, Northeast africa, Shock-metamorphic effects.

INTRODUCTION

The African countries richer in meteorite finds and falls are Morocco, Algeria, Libya, Egypt, Sudan, Tunisia, and Mauritania (Ouknine *et al.*, 2019). However, these meteorites are almost totally exported out of their countries of find by dealers, collectors and foreign scientists. Most classes of meteorites, including Martian (major) and lunar (minor), are found in the African hot deserts.

---

[*] **Corresponding author Mohamed Th. S. Heikal:** Geology Department, Faculty of Science, Tanta University, Tanta, Egypt; E-mail: mohamed.hekal1@science.tanta.edu.eg

Abderrahmane Ibhi, Giorgio S. Senesi, Lahcen Ouknine, Fouad Khiri (Eds.)

The current chapter focuses on the distribution of North East African (Egypt, Sudan and Libya) meteorites (falls and finds) in combination with impact craters if present, and on related phenomena that have caused powerful shock effects. This chapter is very urgent for junior geologists, astronomers, astrophysicists and other researchers.

EGYPT

Egypt is located in the extreme northeastern part of Africa. It occupies about 1,000,000 Km², bordering at the north with the Mediterranean Sea, at east with the Red Sea, at west with Libya and at south with Sudan. The first record of meteorite falls in Egypt was on June 28, 1911, at 9:00 am, when suddenly a number of meteorites slammed on the Egyptian village of El-Nakhla El-Baharia located in the Delta Nile, Abu Hommos region (Figs. **1** and **2**) (Hume, 1911; Ball, 1912; Prior, 1912 and Attia *et al.*, 1955). Recalling the fall site, this meteorite is called Nakhalite. Many investigators have studied and characterized this meteorite (Treiman, 2005; Sautter *et al.*, 2006; Imae and Ikeda, 2007), concluding that it is of Martian type, very rich in olivine and pyroxene. Subsequent studies (Domeneghetti *et al.*, 2013; Alvaro *et al.*, 2015) found that the augite mineral in this meteorite crystallized at 600°C, and was cooled within lava flows at a burial depth of 85 m.

Fig. (1). Map showing the Nile Delta in Egypt. The red circle indicates the El-Nakhla El-Baharia site.

Fig. (2). A piece of the meteorite that fell near El-Nakhla El-Baharia in northern Egypt.

Recently, a number of meteorites were discovered in El-Gilf El-Kebir, which is located in the southwestern parts of the Egyptian Western Desert (Paillou *et al.*, 2004). Most meteorite components are of an iron type and silica glass, therefore, the area where these meteorites were recovered was called Wadi El Crystal (Bewdian name).

The first well-preserved impact crater in Egypt was identified in 2010 by a professional Egyptian-Italian teamwork nearby Gebel Kamil (22°01′ N–26°05′ E) in southern Egypt and was named Kamil Crater (Folco *et al.*, 2010, 2011) (Fig. **3**). The discoverers described it as a bowl-shaped cavity of 45-m diameter and 10-m depth on a Cretaceous sandstone target with a well-preserved ejecta blanket (Figs. **3** and **4**), also displaying some well-developed ejecta rays (Figs. **5** and **6**). These features highlight the exceptional freshness of the structure to which an estimated age between 1600 and 400 BC was attributed (Sighinolfi *et al.*, 2015). A geophysical expedition was undertaken under the leadership of Profs. Folco and El Sharkawy and their teamwork (Fig. **7**) in February 2010 revealed that the crater shows an upraised rim of about 3 m above the pre-impact surface, which is typical of simple craters. The true crater floor depth is about 16-m deep and is overlain by about 6-m thick crater-fill material (Urbini *et al.*, 2012). Cm-scale masses of scoriaceous impact melt glass occur in and close to the crater, which indicates local shock pressures.

Based on geo-archeological supports, namely the chronological relationship between the impact structure and a number of features (trails, settlements) that attest to the prehistoric human occupation of the area, Folco *et al.* (2011) suggested that the impact event occurred less than 5000 years ago. This young age was later confirmed by thermoluminescence dating of quartz in the shocked target rocks, which yielded a formation age between 2000 and 500 BCE, with a favored age interval between 1600 and 400 BCE (Sighinolfi *et al.*, 2015).

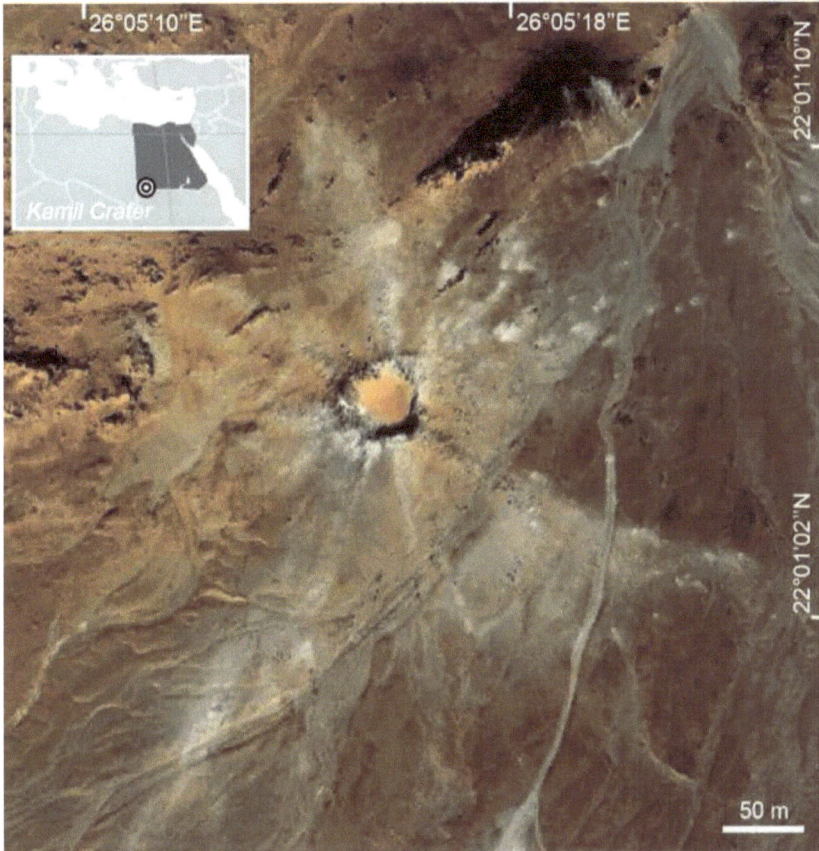

Fig. (3). Location of the Gebel Kamil impact crater. Source (Google Earth).

Fig. (4). Meteor impact crater at Gebel Kamil (Southwestern Desert of Egypt). Source: Photo by El Sharkawy with oral permission.

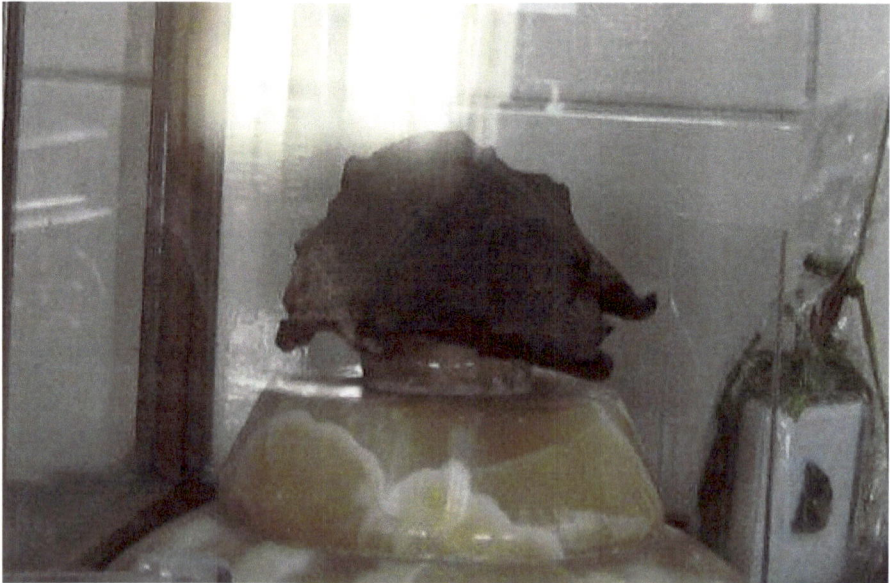

Fig. (5). A sample of the Gebel Kamil meteorite preserved in the Egyptian Geologic Museum, Cairo, Egypt.

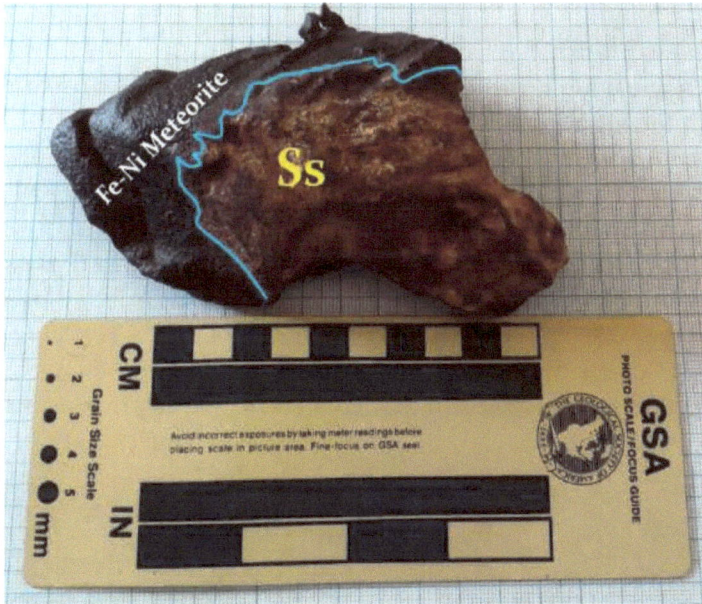

Fig. (6). A sample of the iron-nickel meteorite from the Gebel Kamil impact crater. Ss = sandstone.

Fig. (7). Prof. El Sharkawy with a big sample of the iron-nickel meteorite near the impact crater of Gebel Kamil, Western Desert of Egypt (Source: Folco *et al.*, 2011 and communicated permission with Prof. El Sharkawy).

According to Barakat and El Baz (2001) and Barakat (2011), the Late Proterozoic basement of the central Eastern Desert of Egypt may contain a 7-km diameter eroded impact structure in the locality of Ras Zeidun (25°37′ N-33°52′ E). Barakat (2011) reports the presence of huge concentric plugs of several km in diameter in the structure of unusual jasper deposits. The petrographic investigation carried out by the same author (Barakat, 2011) indicates that the quartz grains in the jasper deposit display characteristics of a meteorite impact, and that the Ras Zeidun site features the basic aspects of eroded multi-ring impact structures. However, no proof for these assessments was presented, and the Ras Zeidun round shape is not clearly visible by satellite images. Moreover, optical and radar remote sensing investigations in the region do not exhibit unusual structural characteristics. However, sentinel 2A principal component analysis (PCA) revealed a change in lithology in the Ras Zeidun region, although no circular structures are observable (Fig. **8**). Therefore, the existence of this allegedly complex impact structure in Ras Zeidun is still controversial.

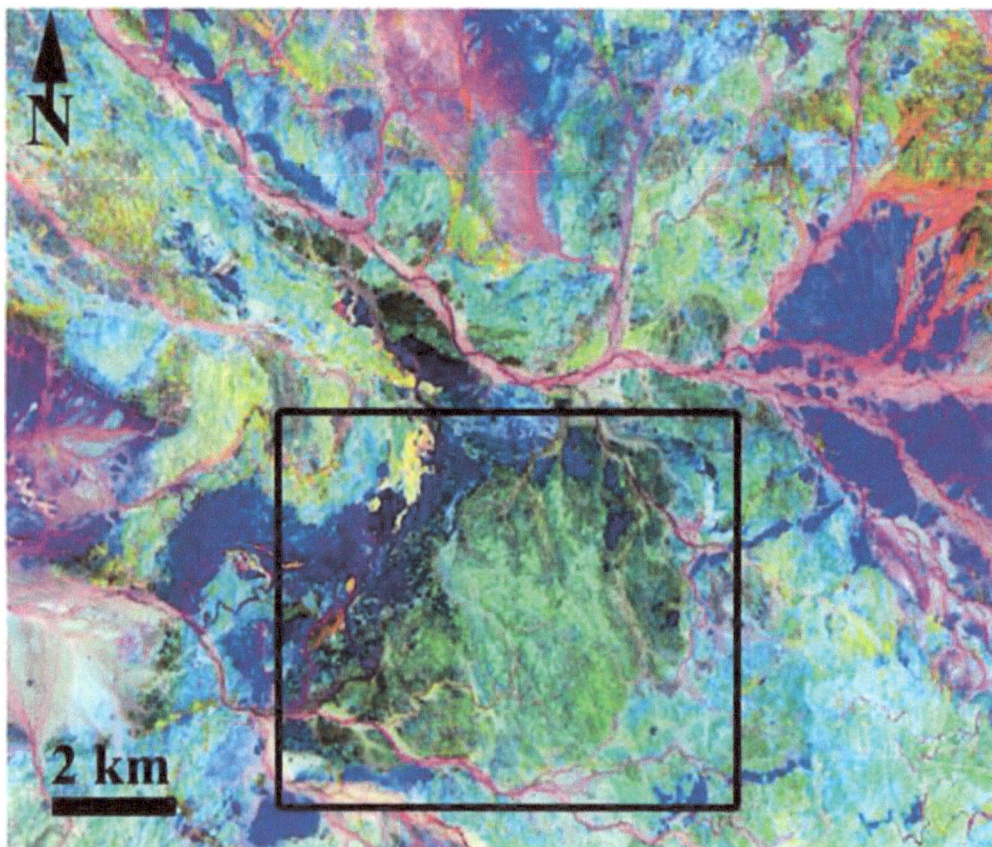

Fig. (8). The black rectangular box represents the PCA image of the alleged Ras Zeidun structure. Source: Barakat, 2011.

SUDAN

Sudan is situated in north-eastern Africa, between Egypt to the north and Eritrea to the south, bordering the Red Sea in northeast, Libya in north west, Chad in west, the Central African Republic in southwest, South Sudan in south and Ethiopia in southeast. Sudan occupies an area of 1,886,068 km², and is the 16th largest country in the world.

Very few studies are reported on meteorites falls and finds in Sudan. Two stones of 60 and 40 kg were found by a geological field trip in 1986 near Kidairat, Northern Kardofan Province, 20 km SE of Umm Badr, WSW of Khartoum, which are said to have fallen in 1983. The type of meteorites seems to be ordinary chondrite (H4), where olivine minerals are predominant. The 40-kg piece is kept at the Geology Department, University of Khartoum, whereas the 60-kg piece remains in Kidairat.

The recovery of fragments of a small asteroid that impacted on northern Sudan on October 7, 2008, was reported by Prof. Shaddad and collaborators of the Physics Department, Khartoum University, and the astronomer Peter Jenniskens (SETI Institute / NASA-Ames) and his students. This is the first asteroid detected prior to impact and observed in space, the fragments of which were recovered in north Sudan. The researchers interviewed eyewitnesses of the fireball, who confirmed that the impact was accompanied by an explosion at an altitude of about 37 km in the atmosphere, with no luminosity below 32 km, which is a behavior more typical of a comet than of an asteroid impact. Never before have meteorites been recovered from an object exploding so high in the atmosphere. Nevertheless, 45 students and their staff of the University of Khartoum were brought to the area just downstream of the explosion location, some 30 km into the desert from the railroad track at Station 6 (Almahata Sitta), a train station between the Sudanese cities of Wadi Halfa and Khartoum, near where the fragments were found (Fig. **9** and **10**). On December 5, after two hours of searching, the first meteorite was recovered by Mohammed Alameen. The next two days, 14 darker and scruffier-looking meteorites were found, which likely belonged to the asteroid 2008 TC3 and gradually increased in size further along the path. A second search in late December raised the total number of recovered fragments up to 47, with a total mass of 3.95 kg and increased diversity of recovered materials. Some of them were very dark and flaky in texture; others were light gray and fine-grained. A third search in late February and early March 2009 measured the vertical extent of the strewnfield, and, thereby, the conditions of ejection during the explosion, raising the total number of recovered fragments to about 300. The meteorite was classified as an anomalous polymict ureilite, now linked to asteroids of class F (written communication with Prof. Shadded, 2016). No impact craters have been detected in Sudan till now, although some circular structures and volcanic craters have been detected, which are similar to meteorites impact craters when viewed from a bird's eye.

Fig. (9). The black fragment of Almahata Sitta meteorite found in the Nubian Desert, northern Sudan (after Peter Jenniskens (SETI Institute/NASA Ames).

Fig. (10). Infrared image of the explosion of the asteroid 2008 TC3 taken by the Meteosat 8 satellite. The path of the asteroid is shown by the yellow arrow; the red-yellow blob on the arrow is the infrared from the explosion.

More recently, in 2015, some geologists found a heavy-weight iron-nickel meteorite (Fig. **11**) in Wadi Halfa at the border between Egypt and Sudan, which is now conserved in the Museum of the Mineral Ministry of Sudan.

Fig. (11). The heavy-weight Fe-Ni meteorite found in Wadi Halfa. Photo by Mohamed Th. S. Heikal (April 2015).

LIBYA

Libya is located in NE Africa and has a surface of 1759540 km^2, *i.e.*, the fourth in size of African countries and the seventeenth in the world. It lies on the Mediterranean sea, between Egypt and Tunisia, with Niger and Chad in the south and Sudan at the southeast.

In Libya, the story of meteorite finds began on December 29, 1932, when Patrick Clayton, a British explorer, discovered a mysterious yellow-green glass scattered across the surface in an area close to western Egypt, at the edge of the Libyan Sand Sea, one of the most remote and inhospitable regions on the Earth. Since then, the Libyan Desert Glass has fascinated scientists who have been puzzled over its formation (Fig. **12**).

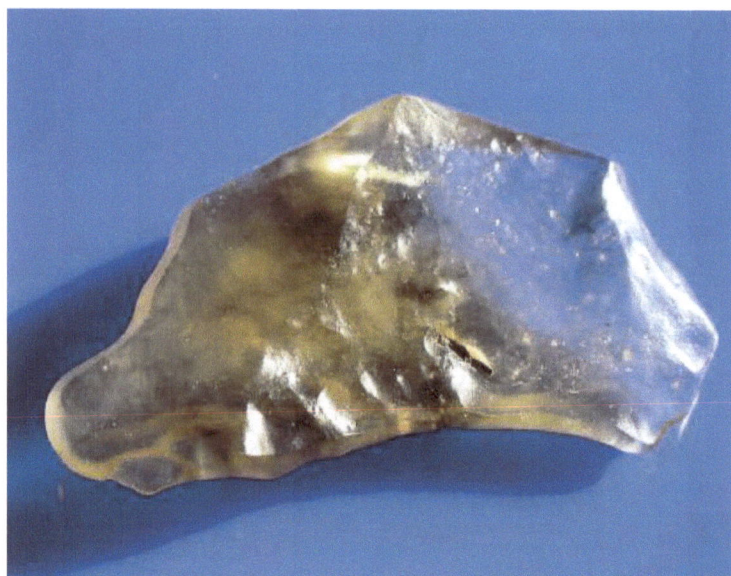

Fig. (12). The Libyan Desert Glass is a natural glass composed of nearly pure silica (Fröhlich *et al.*, 2013).

This meteorite features the purest natural glass in the world, with some remarkable properties. Due to its unusual composition, the formation of this glass has long been considered mysterious. Some of the glass shows inclusions like bubbles, spherules of cristobalite (a high-temperature silica polymorph) and rare dark inclusions, the chemical analyses of which show that they are enriched in meteoritic elements, with typical chondritic proportions. Thus, the best conclusion from these results is that the glass would originate from a meteorite impact on a silica-rich target.

Until July 2001, 1238 meteorites of lunar type have been reported in Libya, most of which were found in two areas, *i.e.*, Dar al Gani and Hamadah al Hamra, which are large, flat and almost featureless desert areas (the so-called "Regs") covered with light-colored sediments, where the erosion rate is low as indicated by the terrestrial age of meteorites (up to 30000 years or even more). Dar al Gani is located on a plateau of marine carbonate rocks with marl components (Fig. **13**), where 869 meteorites (Gibson *et al.* 2011; Van Gasselt *et al.* 2017) have been found from 1990 to 1999 (>800 meteorites from 1994 to 1999), with a relatively low calculated mean recovery density of one meteorite over 6.5 km^2.

In particular, as the preservation of meteorites depends on the combination of specific geological and geomorphological conditions, Dar al Gani represents a perfect site for the recognition and preservation of lunar meteorites (Fig. **14**). This is because of: (i) the absence of quartz sand on the plateau that reduces markedly wind erosion; and (ii) the basic environment emerging from the carbonate ground that retards the rusting of metallic meteorite components. A supposed soil cover during pluvial times has probably protected older meteorites and led to a concentration of meteorites of different periods. The studies of Dar al Gani meteorites suggest the existence of at least 26 strewnfields and 26 meteorite pairs that reduce the number of falls to, at most, 534. However, the use of shock and weathering grades for the recognition of pairings turned out to be problematic, as several strewn fields showed paired meteorites that have been classified as belonging to different shock and weathering grades.

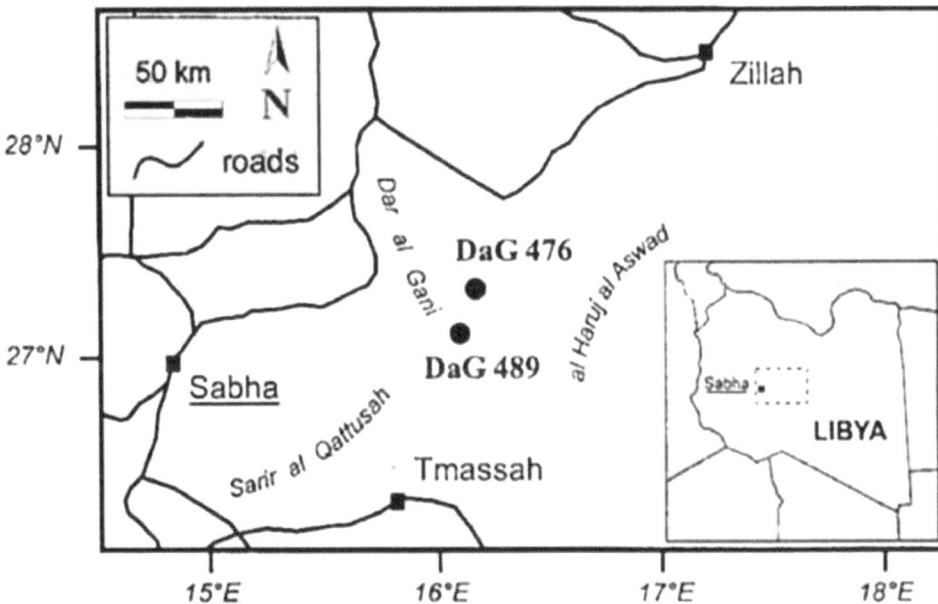

Fig. (13). Location of the Dar al Gani site (Folco *et al.*, 2000).

Fig. (14). Lunar meteorite of anorthositic breccias. The rock appears to be a regolith breccia with at least one large melt-breccia clast (right of center). Photo credit: Randy Korotev.

At present, more than 750 meteorites are classified (Meteoritical Bulletins No. 71, 80-83) based on detailed mineralogical and chemical studies. The classifications include the meteorite class and usually the degrees of shock metamorphism and weathering, the chemical composition of major phases (*e.g.*, olivine and pyroxene) and the striking petrographic features such as brecciation and the occurrence of shock veins or impact melts.

Recently, Hamouda and Alshareeif (2013) analyzed the elemental composition of a meteorite that was recovered in eastern Libya in the Mlodh area located at 39°97′ 0″ N and 7°90′ 0″ E. The authors found that the concentrations of the elements Fe, Na, Ca, and Mg in a terrestrial rock sample were much higher than those found in the meteorite sample. Furthermore, the study revealed that the concentration of rare earth elements (REEs) and Ti in the meteorite sample were much higher than those found in similar terrestrial rocks.

The confirmed impact craters recorded in the Libyan Desert by Chabou (2019) are described below.

– Oasis (24°35′ N–24°24′ E) is an eroded circular structure with a prominent ring of irregular hills that are approximately 100-m high (Abate *et al.*, 1999). The structure lies on Jurassic to Lower Cretaceous sandstones and conglomerates of the Nubia Group underlying Upper Carboniferous clastic sedimentary rocks (Gibson *et al.* 2011). Based on radar images, Koeberl *et al.* (2005a, b) suggested that the structure diameter ranges from 11.5 to 15.18 km, although a more recent study based on high-resolution radar images proposed a diameter of 15.6 ± 0.5 km for this structure (Van Gasselt *et al.*, 2017). Furthermore, French *et al.* (1974) and Gibson *et al.* (2011) discovered many sets of planar deformation patterns in quartz grains of sandstone, which provided evidence of the impact origin of this crater. Although the exact age of the Oasis structure is difficult to be determined, estimates range from 5.5 to 90 Ma (Gibson *et al.* 2011; Van Gasselt *et al.* 2017).

– Jebel Hadid (20°52′N–22°42′E) consists of a circular structure with a diameter of approximately 4.7 km featuring a group of five concentric annular ridges and showing a striking morphological similarity with the Tin Bider impact structure in Algeria. The nearby rocks most likely originate from the Early Cretaceous Nubian sandstone. The Jebel Hadid structure has been studied for the first time by remote sensing imagery that revealed an intricate, deteriorated impact structure that may be dated from the Post-Early Cretaceous to the Pre-Pliocene Period (Schmieder *et al.*, 2009a, b).

– Arkenu Double Crater (Arkenu 1: 22°06′ N–23°45′ E and Arkenu 2: 22°02′ N–23°43′ E), discovered by Paillou *et al.* (2003) in the southeast Libyan desert, in particular, Arkenu 1 measuring 6.8 km in diameter in the northeast, and Arkenu 2, measuring 10.3 km in the southwest. The presence of shatter cone-like structures in the field and impact breccias with planar fractures in quartz grains provided evidence for their origin from a meteorite impact (Paillou *et al.*, 2003). However, this interpretation was disputed by Reimold (2007), Di Martino *et al.* (2008), and Cigolini *et al.* (2012) based on the consideration that the supposed shatter cones would be sandstone wind ablation structures and the reported planar fractures are not actually planar. Furthermore, an extensive study carried out by Cigolini *et al.* (2012) on sample outcropping in the area of the Arkenu structures was unable to find any evidence of shock metamorphism.

– Ibn-Batutah (21°34′ N–20°50′ E), which was discovered by E. M. Ghoneim (2009) using satellite imagery. This crater was formed in the sandstones of Ibn-Battuta that belong to the Cretaceous formation of Nubia, and consists of a bowl-shaped circular structure of a 2.5-km diameter and 25-m depth featuring a round rim raising up to 5 m from the surrounding ground.

CONCLUSION

The first meteorite in NE Africa was discovered in 1911 at Nakhla village in the Middle Delta of Egypt. Meteorite falls and finds in NE Africa, *i.e.*, Egypt, Libya, and Sudan, are comprised of chondrites, achondrites and iron-nickel samples in Egypt and Sudan, and yellow-green glass (Libyan desert glass) and lunar type in Libya. However, Martian and lunar types are fewer in this region. The confirmed impact craters are well-defined at Gebel Kamil, located in the extreme SW of the western desert of Egypt, and various localities in Libya, whereas an undefined impact crater is only mentioned at Wadi Halfa in Sudan.

REFERENCES

Abate, B., Koeberl, C., Kruger, F.J., Underwood, J.R., Jr (1999). BP and Oasis impact structures, Libya, and their relation to Libyan Desert Glass. *Dressler BO, Sharpton VL (eds) Large meteorite impacts and planetary evolution.*, Geological Society of America. *339*, 177-192.
[http://dx.doi.org/10.1130/0-8137-2339-6.177]

Alvaro, M., Domeneghetti, M.C., Fioretti, A.M., Cámara, F., Marinangeli, L. (2015). A new calibration to determine the closure temperatures of Fe-Mg ordering in augite from nakhlites. *Meteorit. Planet. Sci., 50*(3), 499-507.
[http://dx.doi.org/10.1111/maps.12436]

Attia, A., El Shazly, M., El-Shazly, M., Mohrem, O.M., Hozein, A.H. (1955). Meteorites and their equivalent rocks, Geological Museum of Egypt. Arabic Edition. 1, 46.

Ball, J. (1911). The meteorite of El Nakhla El Baharia *Survey Department.*, National printing dept..25.

Barakat, A.A. (2001). El-Baz crater: basaltic intrusion versus meteorite impact crater. *Annals of the Geological Survey of Egypt, 24*, 167-177.

Barakat, A.A. (2011). Ras Zeidun, Eastern desert of Egypt: a possible astrobleme. *Abstracts, 2nd Arab impact cratering and astrogeology conference (AICAC II).*, Hassan II Casablanca University.Casablanca, Morocco10.

Chabou, M.C. (2019). Meteorite Impact Structures in the Arab World: An Overview. *The Geology of the Arab World---An Overview*, Springer.455-506.

Cigolini, C., Di Martino, M., Laiolo, M., Coppola, D., Rossetti, P., Morelli, M. (2012). Endogenous and nonimpact origin of the Arkenu circular structures (al-Kufrah basin—SE Libya). *Meteorit. Planet. Sci., 47*(11), 1772-1788.
[http://dx.doi.org/10.1111/maps.12012]

Di Martino, M., Cigolini, C., Orti, L. (2008). Non-impact origin of the Arkenu craters (Libya). In: Large meteorite impacts and planetary evolution IV conference, Vredefort, South Africa, Lunar and Planetary Institute, Abstract #3012.

Domeneghetti, M., Fioretti, A., Camara, F.M.C., Cammon, C., Alvaro, M. (2013). Thermal history of nakhlites. *Am. Mineral., 90*, 1816-1823.
[http://dx.doi.org/10.2138/am.2005.1773]

Folco, L. (2019). Kamil crater (egypt): a type-structure for small-scale impact craters on ea rth. *Large Meteorite Impacts VI (LPI Contrib. No. 2136).*

Folco, L., Di Martino, M., El Barkooky, A., D'Orazio, M., Lethy, A., Urbini, S., Nicolosi, I., Hafez, M., Cordier, C., van Ginneken, M., Zeoli, A., Radwan, A.M., El Khrepy, S., El Gabry, M., Gomaa, M., Barakat, A.A., Serra, R., El Sharkawi, M. (2010). The Kamil crater in Egypt. *Science, 329*(5993), 804.
[http://dx.doi.org/10.1126/science.1190990] [PMID: 20651117]

Folco, L., Di Martino, M., El Barkooky, A., D'Orazio, M., Lethy, A., Urbini, S., Nicolosi, I., Hafez, M., Cordier, C., van Ginneken, M., Zeoli, A., Radwan, A.M., El Khrepy, S., El Gabry, M., Gomaa, M., Barakat, A.A., Serra, R., El Sharkawi, M. (2011). Kamil Crater (Egypt): Ground truth for small-scale meteorite impacts on Earth. *Geology, 39*(2), 179-182.
[http://dx.doi.org/10.1130/G31624.1]

Folco, L., Franchi, I.A., D'Orazio, M., Rocchi, S., Schultz, L. (2000). A new martian meteorite from the Sahara: The shergottite Dar al Gani 489. *Meteorit. Planet. Sci., 35*(4), 827-839.
[http://dx.doi.org/10.1111/j.1945-5100.2000.tb01466.x]

French, B.M., Underwood, J.R., Jr, Fisk, E.P. (1974). Shock metamorphic features in two meteorite impact structures, Southeastern Libya. *Geol. Soc. Am. Bull., 85*(9), 1425-1428.
[http://dx.doi.org/10.1130/0016-7606(1974)85<1425:SFITMI>2.0.CO;2]

Fröhlich, F., Poupeau, G., Badou, A., Le Bourdonnec, F.X., Sacquin, Y., Dubernet, S., Bardintzeff, J.M., Véran, M., Smith, D.C., Diemer, E. (2013). Libyan Desert Glass: New field and Fourier transform infrared data. *Meteorit. Planet. Sci., 48*(12), 2517-2530.
[http://dx.doi.org/10.1111/maps.12223]

Ghoneim, E.M. (2009). Ibn-Batutah: A possible simple impact structure in southeastern Libya, a remote sensing study. *Geomorphology, 103*(3), 341-350.
[http://dx.doi.org/10.1016/j.geomorph.2008.07.005]

Gibson, R.L., Reimold, W.U., Baegi, M., Crosta, A.P., Shbeli, E., Eshwedi, A. (2011). The Oasis structure, southeastern Libya: new constraints on size, age and mechanism of formation. *42th lunar and planetary science conference.* March 07–11 2011The Woodlands, Texas, USA

Hamouda, S.A., Alshareeif, F.M. (2013). Study of some physical and chemical properties of meteorites in Libya. *International Journal of Astrophysics and Space Science, 1*(2), 7-11.
[http://dx.doi.org/10.11648/j.ijass.20130102.11]

Hume, W.F. First meteorite record.. (1912). Science in Egypt, Cairo Science Journal.

Imae, N., Ikeda, Y. (2007). Petrology of the Miller Range 03346 nakhlite in comparison with the Yamato-000593 nakhlite. *Meteorit. Planet. Sci., 42*(2), 171-184.
[http://dx.doi.org/10.1111/j.1945-5100.2007.tb00225.x]

Koeberl, C., Reimold, W.U., Plescia, J. (2005). BP and Oasis impact structures, Libya: remote sensing and field studies. In: Koeberl, C., Henkel, H., (Eds.), *Impact tectonics, impact studies series 8.* (pp. 161-190). Berlin: Springer.
[http://dx.doi.org/10.1007/3-540-27548-7_6]

Koeberl, C., Reimold, W.U., Cooper, G., Cowan, D., Vincent, P.M. (2005). Aorounga and Gweni Fada impact structures, Chad: Remote sensing, petrography, and geochemistry of target rocks. *Meteorit. Planet. Sci., 40*(9-10), 1455-1471.
[http://dx.doi.org/10.1111/j.1945-5100.2005.tb00412.x]

Ouknine, L., Khiri, F., Ibhi, A., Heikal, M Th.S., Gerantc, T., Medjkane, M. (2019). Insight into African meteorite finds: Typology, mass distribution and weathering process. *J. of Afri. Earth Sci., 158*, 103551.
[http://dx.doi.org/10.1016/j.jafrearsci.2019.103551]

Paillou, P., El Barkooky, A., Barakat, A., Malézieux, J.M., Reynard, B., Dejax, J., Heggy, E. (2004). Discovery of the largest impact crater field on Earth in the Gilf Kebir region, Egypt. *C. R. Geosci., 336*(16), 1491-1500.
[http://dx.doi.org/10.1016/j.crte.2004.09.010]

Paillou, P., Rosenqvist, A., Malézieux, J.M., Reynard, B., Farr, T., Heggy, E. (2003). Discovery of a double impact crater in Libya: the astrobleme of Arkenu. *C. R. Geosci., 335*(15), 1059-1069.
[http://dx.doi.org/10.1016/j.crte.2003.09.008]

Prior, G.T. (1912). The meteoric stones of El Nakhla El Baharia (Egypt). *Mineral. Mag., 16*, 274-281.

[http://dx.doi.org/10.1180/minmag.1912.016.76.04]

Reimold, W.U. (2007). The Impact Crater Bandwagon (Some problems with the terrestrial impact cratering record). *Meteorit. Planet. Sci., 42*(9), 1467-1472.
[http://dx.doi.org/10.1111/j.1945-5100.2007.tb00585.x]

Sautter, V., Jambon, A., Boudouma, O. (2006). Cl-amphibole in the nakhlite MIL 03346: Evidence for sediment contamination in a Martian meteorite. *Earth Planet. Sci. Lett., 252*(1-2), 45-55.
[http://dx.doi.org/10.1016/j.epsl.2006.09.024]

Schlüter, J., Schultz, L., Thiedig, F., Al-Mahdi, B. O., Abu Aghreb, A. E. (2002). The Dar al Gani meteorite field (Libyan Sahara): Geological setting, pairing of meteorites, and recovery density. *Meteoritics & Planetary Science.* 37(8), (1079), 1093.
[http://dx.doi.org/10791093.Doi:10.1111/j.19455100. 2002.tb00879.x]

Schmieder, M., Buchner, E., Le Heron, D.P. (2009). The Jebel Hadid structure (Al Kufrah Basin, SE Libya)—A possible impact structure and potential hydrocarbon trap? *Mar. Pet. Geol., 26*(3), 310-318.
[http://dx.doi.org/10.1016/j.marpetgeo.2008.04.003]

Schmieder, M., Buchner, E., Hofmann, B.A., Gnos, E. (2009). The ash Shutbah circular feature—A suspected meteorite impact site in Saudi Arabia. *Abstracts, the first Arab impact cratering and astrogeology conference,* Amman, Jordan 84-85.

Sighinolfi, G.P., Sibilia, E., Contini, G., Martini, M. (2015). Thermoluminescence dating of the Kamil impact crater (Egypt). *Meteorit. Planet. Sci., 50*(2), 204-213.
[http://dx.doi.org/10.1111/maps.12417]

Treiman, A.H. (2005). The nakhlite meteorites: Augite-rich igneous rocks from Mars. *Chem. Erde, 65*(3), 203-270.
[http://dx.doi.org/10.1016/j.chemer.2005.01.004]

Urbini, S., Nicolosi, I., Zeoli, A., El Khrepy, S., Lethy, A., Hafez, M., El Gabry, M., El Barkooky, A., Barakat, A., Gomaa, M., Radwan, A.M., El Sharkawi, M., D'orazio, M., Folco, L. (2012). Geological and geophysical investigation of Kamil crater, Egypt. *Meteorit. Planet. Sci., 47*(11), 1842-1868.
[http://dx.doi.org/10.1180/minmag.1912.016.76.04]

Van Gasselt, S., Kim, J.R., Choi, Y.S., Kim, J. (2017). The Oasis impact structure, Libya: geological characteristics from ALOS PALSAR-2 data interpretation. *Earth Planets Space, 69*(1), 12.
[http://dx.doi.org/10.1186/s40623-017-0620-8]

CONCLUSION

Africa is a suitable continent for meteorite recovery as the total collected number (falls and finds) represents more than 1/6 of all extraterrestrial rocks collected on the Earth. The high number of meteorite recoveries in Africa is favored by its large surface area, its geographical location between the northern and southern hemispheres, and the wide diversity of climate and natural environments (deserts, forests, mountains).

Since 1800, the number of meteorite falls in Africa has continued to grow, with 80% of them recorded during the period 1910-2014. At the end of August 2020, twenty-one official observed meteorite falls were recorded only in Morocco. The meteorite falls recovery rate during the past 88 years (from 1932 to 2020), for a surface area of 712.5 km^2, is low on average, *i.e.*, approximately four fall recovery every 10-years, equivalent to 0.4 fall per year per 2.11 km^2. The average rate of falls is low in Africa, with only 0.023 per million km^2 per year, but it is twice that recorded in Australia. The classification of meteorite falls in Africa shows an abundance of chondritic falls (76%) and a deficit of achondritic ones (17%). In particular, iron meteorite falls are relatively rare globally, accounting for only 5.9% of the fall population in Africa, which include three Martian meteorites (Nakhla from Egypt, Tissint from Morocco, and Zagami from Nigeria), but no Lunar meteorites. Moreover, the spatial distribution of recorded meteorite falls reveals that most of them were recovered in the Northwest, East, and South of the continent, whose countries combine a number of favorable human and geographical factors, *i.e.*, a large population with a uniform distribution, a local population giving importance to the observation of meteorite falls, a large area with a low percentage of forest cover, a semi-arid to arid climate, and access to the place of fall.

Over 300 years (1716-2017), 9660 meteorites have been discovered in Africa, with 3.18 discovered per 10,000 km^2. Of them, 2399 samples (25%) were found in 23 countries and bear the name of the official collection site, and 7261 meteorites (75%) are without filiation (6781 under the acronym Northwest Africa "NWA", 476 called "Sahara", and 5 called Northeast Africa "NEA"). The temporal evolution of African meteorite finds shows that most of them were collected in the early 21st century, which can be mostly attributed to the expertise acquired by the rural population of the hot deserts. In particular, the rate of meteorite finds does not seem to depend on the population density or its distribution, but is related prevalently to a category of the population, *i.e.*, nomads, passionate people, experienced searchers, scientists, *etc.*, interested in meteorites and endowed with a specific skill necessary to access privileged hunting grounds and distinguish extraterrestrial rocks from terrestrial rocks.

Currently, fragments from a total of 1747 distinct and authenticated meteorites from Morocco with a total mass of 6175 kg have been described. Most of these meteorites were found in the Tata, Zogora, Errachidia, Es Smara and Erfoud regions of Eastern and Southern Morocco. To date, the largest quantity of a single meteorite fall in Morocco remains that of Agoudal iron found in 2012 on the Imilchil region in Moroccan Haut Atlas, with several tons distributed in samples from a few grams to several hundred kilograms.

The meteorite classification of African discoveries indicates that this collection includes all types of meteorites known to date, and is characterized by an abundance of stony meteorites with significant rates of rare meteorites. Over the past three decades, African discoveries have seen a decrease in chondrites, especially ordinary chondrites, and a significant increase in the number of carbonaceous chondrites and achondrites. This result can be explained by economic reasons, *i.e.*, the increase in prices of carbonaceous chondrites and achondrites, and by the non-declaration of ordinary chondrites to the Nomenclature Committee. Similar to Australian and Antarctic collections, the finds in Africa feature a low abundance of metallic meteorites, which may be mostly explained by the recovery and use of this type of meteorite in prehistoric and historic times by Northern African civilizations.

The distribution of sample masses of African finds shows that many meteorites (54%) have average masses between 100 and 1000 g. Most small meteorites are concentrated in the northern hemisphere, while the majority of the heaviest ones have been collected in the southern hemisphere. The collection of African meteorites shows a distribution of masses typical of areas where the search for meteorites is carried out on foot by nomads or by car by meteorite hunters, especially in the desert areas of Morocco, Algeria and Libya.

African meteorites show weathering grades ranging from W0 to W6, with most samples (66%) featuring a weathering degree lower than W3, which proves their better preservation. The factors that influence the mechanisms and the degrees of alteration of African meteorites include: (i) *Climate*. This is the main factor allowing the preservation and accumulation of meteorites in North (Sahara desert) and South-West Africa (Namib and Kalahari deserts) thanks to favorable semi-arid to arid conditions, whereas in the tropical zones and in the southeast of the continent, meteorites disappear because of the warm and humid climate; (ii) *Mass of the sample*. The existence of an apparent correlation between the masses of meteorites and the degree of their alteration reveals that small samples are altered more than big ones; (iii) *Terrestrial age of the sample*. The apparent correlation existing between the degree of weathering and the terrestrial age of the sample shows that the African finds collection is less altered due to its relatively recent age (80% of the samples have terrestrial ages below 20 kyrs); and (iv) *Porosity of the sample*. The collection comprises 49% of samples featuring a shock level between S0 and S2 and 51% having shock levels between S3 and S6. Using this parameter as an indication of the initial porosity, meteorites with higher initial porosity (S0 - S2) result generally more weathered than meteorites with lower initial porosity (S3 - S6). Thus, the initial porosity is an important factor controlling the weathering process of meteorite finds in Africa.

In general, African meteorites are dominated by low altered stony meteorites, almost all of which (99%) are recovered from the continent hot deserts that allow their preservation and accumulation, and simplify their search. This explains why the African continent is one of the most prolific sites for meteorite recovery in the world.

After the adoption of the appellation "NWA meteorite" in 2000, almost all meteorites found in Northwest Africa are declared under this acronym. In particular, most NWA meteorites collected until January 1st, 2015, originating from Morocco, and only 23% of them have no information on their country of origin. The lack of information noted by the Nomenclature Committee on this population of meteorites can be explained by: (i) the strict legal and regulatory measures (especially in Algeria), (ii) the modest socio-cultural and socio-economic conditions that characterize certain areas of the region, and (iii) the security disturbances that have reigned in Northwest Africa (Sahelian belt) over the past decade.

The typology of the NWA collection shows the predominance of stony meteorites (97%) and the presence of mixed and metallic meteorites with identical proportions (1.5%). However, rare types of significant scientific value are included. To highlight this collection, a new database of NWA meteorites has been developed, and a re-nomenclature of NWA meteorites has been proposed, according to the guidelines of the new Guide to meteorite nomenclature. This action is believed to enhance the inestimable scientific and socio-economic value of the meteoritic heritage of Morocco and other countries of Northwest Africa.

In particular, meteorites found in Morocco are playing an increasingly important role in fundamental research across a wide spectrum of disciplines, with new groups of meteorites being recovered that are extending our knowledge of the early Solar System. The astronomical representations engraved on the rocks would confirm the ancient concern in observing meteors and meteorites. In particular, three petroglyphs found near Ida Ou Kazzou (approximately 100 km north of Agadir) would suggest that ancient Moroccans observed meteorite falls.

The record of meteorites impact craters in North Eastern Africa is still incomplete and debated. Of the 190 confirmed impact structures/sites on the Earth crust, only less than 8 have been identified in North Eastern Africa, including Oasis in Libya and Kamal in Egypt. The only criterium that provides evidence for an impact origin of circular geological structures is the occurrence of shock-metamorphic effects, *i.e.*, the presence of shatter cones and diagnostic shock-metamorphic mineral deformation and transformation phenomena, in the target rocks. These effects are evident in the Kamal crater (Egypt) and most impact craters in Libya, whereas a few other structures of alleged impact origin are still controversial. The discovery of preserved meteorite fragments and the detection of the geochemical signature of meteoritic indicators should be considered very carefully during fieldwork and further advanced studies.

In Morocco, two meteorites finds could be associated with impact craters: (i) 300 kg of mesosiderite fragments were collected from an area of 50 x 100 m elongated roughly along the southwest to northeast axis (N30) in the Toufassour region in the year 2003, and (ii) after the first fragment was collected by a nomad in 2007, more than a ton of specimens weighing 100 g to more than 196 kg were harvested over a period of 6 years in the Agoudal iron meteorite strewnfield. Some sources speculate that even a still bigger quantity (up to 2 tons) has been recovered. The Agoudal area is a unique site in Morocco where the relics of several possible impact structures have been discovered.

In the Sahara Desert, a very relevant number of well-preserved meteorites has been collected, as their conservation was facilitated by the desert dry climate in which a meteorite would require centuries to degrade. The recovery of meteorites as quickly as possible after their fall guarantees their preservation and provides a scientific material of quality that allows to infer questions about the conditions of their formation and development prior to fall and residence. However, the Sahara Desert continues to provide findings from large meteorite falls several years after the first discovery, but the first and last recoveries seem not to be the same. The physical aspect and sometimes some chemical parameters are different, which is due to the alteration that meteorites undergo over the years.

In conclusion, the North African region, *i.e.*, the Sahara desert, has become one of the most important places to recover meteorites, together with Antarctica, Atacama and the great deserts of North America. Nevertheless, it is estimated that more than 90% of the desert surface has not yet been explored, and it is still expected to hide a huge number of meteorites. New finds are classified annually, and countries such as Mauritania, Mali and Egypt are emerging in this, attracting competition. The most optimistic forecasts suggest that many new meteorites will continue to be recovered from the great Sahara Desert in the coming decades. Preserving them properly will be essential to provide science with such important objects.

SUBJECT INDEX

A

Ablation 5, 33, 75, 148
  atmospheric 33
Achondrites 12, 13, 19, 20, 21, 34, 35, 36, 51,
    67, 68, 69, 105, 106, 199
  asteroidal 20
Acid, nitric 9, 23
Acronym NWA, generic 107
Ad-hoc assumptions 125
African 29, 30, 31, 33, 41, 43, 48, 49, 53, 55,
    62, 67, 69, 74, 75, 77, 79, 82, 83, 181,
    199, 200
  climates 77
  continent 30, 31, 33, 43, 48, 67, 83, 200
  deserts 75
  discoveries 199
  distribution 79
  educational systems 55
  hot deserts 181
  initiative 62
  population 33, 41, 49, 53, 69, 74, 82
  territory 29, 48
African countries 42, 43, 44, 45, 46, 50, 52,
    53, 54, 55, 59, 60, 62, 63
  colonization of 45, 60
African meteorite 58, 59, 67, 79, 82
  features 59
  population 58, 67, 79, 82
Agoudal 143, 145, 201
  area 201
  Imilchil meteorite 145
Analytical techniques 143
Ancient 115, 116, 119, 122, 201
  cultures 115, 119
  Egyptians 122
  Moroccans 116, 201
Antarctic 68, 74, 77, 78
  and Australian populations 77
  meteorite populations 74
  meteorites 68, 74, 78
Antarctica 58, 177

  and polar areas 177
  deserts 58
Artist engraving technique 116
Asian monsoons 75
Asteroids 2, 3, 4, 5, 7, 15, 16, 20, 21, 188, 189
Astromythology 115
Astronomy 46, 62
Atacama collection 78
Australian 51, 71, 79
  mass frequencies 71
  meteorites 51, 79

B

Breccias 22, 131, 143, 146, 147
  fine-grained monomict 131
  regolithic 22
  volcanic 143
Butler meteorite 150

C

Cameras 7, 8, 50, 62, 177
  automatic 8
  portable 62
  spectroscopic video 50
Carbonaceous chondrites (CC) 10, 15, 16, 69,
    77, 82, 105, 106, 113, 128, 199
Chinguetti 103
Chlorapatite 143
Chondrite 20, 80
  meteorite matrix 80
  chemistry 20
Chromite 131, 132, 134, 136, 143, 145
Chronology 118, 139
Circular geological features 181
Climate 49, 113, 199
  humid 113, 199
  subtropical 49
Clinopyroxene 21, 22, 129, 131, 145
  crystals 145

T

Thermal metamorphism 16, 18, 20
Toufassour depression 145
Transported meteorites 97, 98

W

Weather bulletin database 31
Weathering 58, 73, 74, 75, 77, 78, 79, 80, 81,
    82, 123, 127, 128, 130, 133, 137, 168,
    192, 193, 200
  chemical 75, 79
  effects 78, 123
  factors 58, 73
  grades 73, 74, 75, 77, 78, 79, 80, 81, 82,
    127, 130, 192
  process of meteorites 77, 200

www.ingramcontent.com/pod-product-compliance
Lightning Source LLC
Chambersburg PA
CBHW050839220326